Leo Lesquereux

Contributions to the fossil flora of the western territories

Part 3

Leo Lesquereux

Contributions to the fossil flora of the western territories
Part 3

ISBN/EAN: 9783337268688

Printed in Europe, USA, Canada, Australia, Japan

Cover: Foto ©berggeist007 / pixelio.de

More available books at **www.hansebooks.com**

DEPARTMENT OF THE INTERIOR.

REPORT

OF THE

UNITED STATES GEOLOGICAL SURVEY

OF

THE TERRITORIES.

F. V. HAYDEN,

UNITED STATES GEOLOGIST-IN-CHARGE.

VOLUME VIII.

WASHINGTON:
GOVERNMENT PRINTING OFFICE.
1883.

NOTE.

DEPARTMENT OF THE INTERIOR,
UNITED STATES GEOLOGICAL SURVEY.
November 1, 1883.

On the 27th of September, 1882, at the request of Dr. F. V. Hayden, the completion of the publications of the United States Geological and Geographical Survey of the Territories, formerly under his charge, was committed to the charge of the Director of the Geological Survey by the following order from the honorable the Secretary of the Interior:

DEPARTMENT OF THE INTERIOR,
Washington, September 27, 1882.

Maj. J. W. POWELL.
Director U. S. Geological Survey, City:

SIR: The letter of Prof. F. V. Hayden, dated June 27, bearing your indorsement of July 20, relating to the unpublished reports of the survey formerly under his charge, is herewith returned.

You will please take charge of the publications referred to in the same, in accordance with the suggestions made by Professor Hayden.

It is the desire of this office that these volumes shall be completed and published as early as practicable.

Very respectfully,

H. M. TELLER,
Secretary.

Of the publications thus placed in charge of the Director of the Survey, the accompanying volume is the second to be issued. The first was entitled "The Vertebrata of the Tertiary Formations of the West, by Edward D. Cope." On the 12th day of October, 1882, the manuscript of the present volume was received at the office of the Geological Survey, and through the hearty co-operation of Professor Lesquereux, the work has been pushed to rapid completion. The volume is an important contribution to the ancient botany of North America, and will be heartily welcomed by paleontologists.

Director.

LETTER TO THE SECRETARY.

WASHINGTON, *November 1, 1883.*

SIR: I have the honor to transmit, for your approval, the eighth volume of the final reports of the United States Geological and Geographical Survey of the Territories, prepared by the eminent paleontologist, Prof. Leo Lesquereux.

A brief synopsis of the contents of the volume may be given as follows:

In the first part—the Cretaceous Flora—are described a large number of new species, some representing rare and very remarkable types, all of which are figured on the first seventeen plates. Besides the description of the species, there are some general remarks on the geology of the Dakota group, and on the character of the plants in regard to climate and their affinities with plants of succeeding geological periods. A table of distribution is added, enumerating all the species known up to the present time, pointing out the relations of the plants of Europe and various parts of North America with those of the Dakota group in Nebraska, Kansas, and Colorado. The number of species enumerated in this table is 443, of which 200 are from the Dakota group.

The second part contains a revision of the plants of the Laramie group. The introduction considers the relations of these plants to those of Europe, for the purpose of fixing the age of the formation. Then follows a description of a few new species from very fine specimens on three plates, and a table of distribution including only the species of the Laramie group, which in the seventh volume of the series were mixed with those of the other stages of the Tertiary and were not grouped clearly enough for the proper appreciation of the general characters of the flora.

Up to the present time the author has been unable to find a single species that he could identify with any from the Dakota group. He has now in his possession very large collections of plants from this group, which have not been reported upon, collected in Colorado and Wyoming; yet after a careful examination he fails to find any form even related to those of the Dakota group.

The third part reviews the flora of the White and Green River regions, which he separates into two groups. The plants of Green River and Alkali

Stations and Randolph County, Utah, are most of them different from those
of Florissant, Mouth of White River, and Elko. These plants are represented by twenty-one plates, and their relation is indicated with the flora of
the Gypses of Aix in France, which is generally regarded as lowest Miocene
or Oligocene. The table of distribution of these plants includes, in America,
those of Florissant, Elko, Green River Station, Alkali Station, Sage Creek,
and Barrell Springs as compared with the Miocene of Greenland, Alaska,
the Oligocene of France and Germany, and the Miocene of Europe.

The fourth part relates to Miocene plants described from specimens
obtained from the Bad Lands, California, and Oregon, and from Alaska,
and they occupy fifteen plates. There is also a table of distribution that
indicates the relations of these species of Alaska, Carbon, Washakie, the
Bad Lands, Oregon, California, and Fort Union with the Arctic Miocene,
Greenland, Spitzbergen, and those of Europe. This eighth volume forms
a kind of supplement to the two preceding volumes, inasmuch as in it are
figured and enumerated all the plants which have been found since their
publication, in the formations of the Mesozoic and Cenozoic periods of
North America, and therefore forms a broad basis in vegetable paleontology for the direction of future researches and the classification and
determination of the fossil flora of the Continent. The three volumes of
this series, on vegetable paleontology, form a grand monument to the
industry and fame of the author.

I take pleasure in acknowledging my obligations to the Director of the
U. S. Geological Survey, who has with great kindness superintended the
printing of this Report.

The plates were engraved by the well-known firm of Thomas Sinclair
& Son, of Philadelphia, and are fine examples of their work.

I have the honor to remain, with great respect, your obedient servant,

F. V. HAYDEN,
United States Geologist.

To the Honorable the SECRETARY OF THE INTERIOR.

CONTRIBUTIONS

to

THE FOSSIL FLORA

OF THE

WESTERN TERRITORIES.

PART III.

THE CRETACEOUS AND TERTIARY FLORAS.

By LEO LESQUEREUX.

WASHINGTON:
GOVERNMENT PRINTING OFFICE.
1883.

CONTENTS.

LIST OF ILLUSTRATIONS.

LETTER OF TRANSMITTAL.

COLUMBUS. OHIO, *September 30, 1882.*

Dr. F. V. HAYDEN, *Philadelphia.*

DEAR SIR: I send herewith the manuscript of the eighth volume of the Reports of the United States Geological Survey of the Territories. made under your direction.   Besides a short introduction, this volume contains:

1st. A review of the Cretaceous Flora of the Dakota Group, or of what has been published in volume VI, with descriptions of a large number of new and remarkably interesting species illustrated by 17 plates.

2d. Some remarks on the Flora of the Laramie Group, which I consider as Eocene, with descriptions of a few new species, illustrated by 3 plates.

3d. The more valuable part of the volume, viz: the descriptions of the plants of the Oligocene, a flora of which little was known before, and which is now richly represented by a large number of specimens, especially from Florissant, Colorado.   This Flora will be quite as well received by paleontologists as has been the Cretaceous Flora of volume VI.   It is illustrated by 24½ plates, which are all very finely made.

4th. Half of one plate serves for illustrations of a few plants from the oldest Pliocene, or upper Miocene of California.

5th. Descriptions with figures of Miocene plants of the Bad Lands, with 5 plates.   The plants, clearly of Miocene type, are very interesting from their relation to species of the Arctic Flora.

6th. Descriptions of species of Miocene plants of California and Oregon from specimens pertaining to the State Museum of Oakland, California. They are illustrated by 10 plates, the whole number of the plates being 60.

7th. A short account and description of new species found in a collection of fossil plants made in Alaska by W. H. Dall, of the United States

Coast Survey, for the Smithsonian Institution. The specimens were sent to me for determination, and I was allowed to give in volume VIII a short description of the new species added to the Alaskan Flora already partly known by the works of Heer. These new species have been figured in the Proceedings of the National Museum, vol. v, pl. vi–x.

It is not unnecessary to remark that all the plants described in volume VIII are considered in separate groups according to their relation to the age of the formation which they determine. Comparisons are established with the European Floras by tables of distribution, etc.

I truly believe that this volume will prove to be a very valuable contribution, not merely to the paleontology but also to the geology of this country.

Very truly and respectfully yours,

LEO LESQUEREUX.

CRETACEOUS AND TERTIARY FLORA.

BY LEO LESQUEREUX.

INTRODUCTION.

The present volume contains:

1st. The materials referable to the Cretaceous Flora.

The species recognized from specimens received since the publication of the Annual Report of Dr. F. V. Hayden, 1874, are of course described here, but it has been found advisable to add to them and to consider again part of what has been published in that report as a Review of the Cretaceous Flora of North America; mentioning also the species described by Professor Heer and Dr. Newberry from specimens obtained from the Dakota Group.

It is well known that the plants of the Cretaceous epoch, at least those of a higher class, the Dycotyledons, have been barely discovered and described in Europe, while the profusion of these vegetables in the Dakota Group constitutes an original illustration of a peculiar vegetation which, for reasons explained hereafter, will be of great significance in the future. From this consideration the exposition, in the same work, of all that is known to this time of the North American Cretaceous Flora is greatly to the advantage of vegetable paleontology both in this country and in Europe.

2d. A description of a few species of plants of the Laramie Group, which I persist in considering as Eocene.

These species, added in this volume to the list of the plants already described from the same formation, were all obtained at Golden, Colorado, from the locality where most of those published formerly were found by myself. One, *Oreodoxites plicatus*, a fine Palm, represented by a number of well-preserved though more or less fragmentary leaves, is of a peculiar type, and finds its affinity only in *Ludoviopsis geonomæfolia*, Sap., of the Eocene of Sézanne. A second, *Sterculia modesta*, Sap., also of Sézanne, is represented by a beautifully preserved specimen whose identity has been recognized by the author. A third, *Aralia pungens*, is remarkable for its very close relation, perhaps identity, to four species described by Massalongo as *Sylphidium* from the Eocene of Italy. And still a fourth, *Zizyphus Beckwithii*, is evidently allied to *Z. Harcourtii* of Sézanne. These, on seven species only, added to the flora of the Laramie Group, tend to confirm the conclusions which I have admitted on the age of the flora of the great Lignitic, or Laramie, Group.

3d. A large number of species described from what I called in Volume VII the Green River Group No. 4, which I considered as probably Miocene.

When that volume was published this flora was known only by a very few species. Since that time a large number of specimens have been procured from the same formation, especially at Florissant, Colorado. The species which they represent are very interesting as indicative of a geological period older than the Miocene, or preceding in age the Carbon and Alaska floras.

4th. A new contribution to the Miocene Flora from specimens procured from various localities of the Bad Lands of California and Oregon, with mention of new species recently obtained from Alaska, and a note upon a few specimens from the Chalk Bluff of California, a Pliocene formation.

I.—THE FLORA OF THE DAKOTA GROUP.

GENERAL REMARKS.

All that refers to the geology of the Cretaceous Dakota Group—its immediate superposition upon rocks of Permian age; its relation to the strata overlying it in an uninterrupted series of marine deposits up to the base of the Tertiary measures; its thickness, the superficial expanse of its area—has been recorded in the general remarks of Volume VI of these reports. Since that time very little has been added to what was known and published on the subject.

One fact only should be mentioned now. It is the discovery of numerous specimens of Cretaceous plants at the base of the Rocky Mountains in

Colorado. The plants, by the identity of a number of them and the close affinity of character of some others with species of the Dakota Group, have positively confirmed the supposition that this formation, passing westward in Kansas under the Tertiary measures, is prolonged under them and continues to the Rocky Mountains.

Already, in 1873, Dr. A. C. Peale had procured from Colorado fragments of poorly preserved leaves which had been recognized as identical with *Proteoides acuta*, Heer, a species commonly found in the Dakota Group of Kansas and Nebraska. From this, Nos. 14–16 of the section of South Platte River[1] had been then considered by Dr. Hayden as referable to a Cretaceous formation. More recently, Passed Assistant Engineer H. C. Beckwith, United States Navy, and Rev. Arthur Lakes, have got, near Morrison, a few miles west of Denver, numerous specimens of some of the more predominant species of the Dakota Group—*Sassafras* (*Araliopsis*) *cretaceum*, *Magnolia Capellini*, *Aralia*, *Salix protecefolia*, etc., with some others, which though new are related species which tend to identify the Cretaceous formation at the base of the Rocky Mountains with that of Kansas. Admitting, therefore, the prolongation of the Dakota Group under the Tertiary measures to the base of the mountains, the width of the area covered by this formation should be estimated from east to west at 450 to 500 miles.

Perhaps, also, I should omit here any remarks on the flora of the North American Cretaceous as represented by the plants of the Dakota Group, having already, in Volume VI of the United States Geological Survey of the Territories, by Dr. F. V. Hayden, considered the general character of this flora and its relation to plants living at our time, or to analogous or identical species observed in the formations succeeding that of the Cretaceous. But the materials which I had then for consideration were few and local; they have since been greatly increased, and also new points for comparison have been furnished to phytopaleontologists by the works of Heer on the recently discovered Cretaceous plants of Greenland. From this, some of the conclusions formerly admitted have been more or less modified, while others have received a higher degree of precision

---

[1] Dr. F. V. HAYDEN, *Annual Report*, 1873, pp. 195, 196.

if not of actuality.   It is thus advisable to look again over what is known
to the present time of the characters of the North American Cretaceous
flora and to record the deductions legitimately derived from that knowledge.
This kind of work is a necessity for the present, as it will be also for the
future, not only because what is known now is, probably at least, a mere
fraction of the elements constituting the North American Cretaceous flora,
but because the determinations of the plants are still and must be for a
long time to come unreliable to a certain degree.

   The plants of the Dakota Group, as known mostly by detached leaves,
are striking from the beauty, the elegance, the variety of their forms, and
from their size.   In all this they are fully comparable to those of any geological epoch as well as to those of our time.   From entirely developed
leaves, less than one inch in size, they show all the gradations of size to
one foot, even to a foot and a half in diameter.   The multiplicity of forms
recognized for a single species is quite as marked as it might be upon any
tree of our forests; and to show the admirable elegance of their forms
it suffices to say that, at first sight, they forcibly recall those of the most
admired species of our time—the Tulip-tree, the Magnolia, the Sassafras,
the Sweet-gum, the Plane-tree, the Beech, the Aralia, etc.   The leaves of
Protophyllum Sternbergii have the size and the aspect of those of the Catalpa.
one of our finest ornamental trees.   Those of *Menispermites obtusilobus*, of
Protospermum quadratum, represent in the same manner some of the rarest
shrubs, *Menispermum*, *Ferdinandia*, etc., carefully raised in conservatories
for the graceful forms of their leaves or the richness of their vegetation.
It is, indeed, the first impression received from the beauty of forms of the
leaves of the North American Cretaceous, and the evident likeness of their
facies to that of the finest vegetable types of our time, as we see them around
us, which strikes the paleontologist, and may lead him into error in forcing
upon the mind the belief of a typical identity where possibly there may be a
mere likeness of outlines, a casual similarity of forms in the leaves.   For,
really, when we enter into a more detailed analysis of these Cretaceous
leaves, we are by and by forcibly impressed by the strangeness of the characters of some of them, which seem at variance with any of those recognized
anywhere in the floras of our time, and unobserved also in those of the
geological intermediate periods.   Not less surprised are we to see united in

a single leaf, or species, characters which are now generally found separated in far distant families of plants. The leaves of *Eremophyllum*, so striking by the peculiar appendages of their borders; those of *Anomophyllum*, referable to *Platanus* by one-half. to *Quercus* by the other; those of *Platanus obtusiloba*, half *Acer*. etc.. are of this kind.

On another side, the characters of some of the Cretaceous species are sometimes of such a transient or indefinite order that it is scarcely possible to take hold of them and to describe them with any degree of reliance. At first sight they appear very distinct, but, in comparing a number of specimens, the differences dwindle by unmistakable transitions and disappear. In other leaves, on the contrary, visibly identical by their outlines. the nervation is so different that they are forcibly separated and referred to far distant generic divisions. Hence this flora does not leave any satisfaction, any rest. to the mind. Even the most clearly defined types become doubtful in regard to their integrity when we see others, which. at first, were recognized as positively fixed, manifesting instability and pointing to diversity of relation by the discovery of new specimens. The leaves considered first as *Sassafras*, for example, seemed evidently referable to this genus; but when leaves of the same type were found with dentate borders. though bearing. besides, all the characters of a genus which belongs to the *Laurineæ*, a family where, as yet, no representative has been found with dentate borders of leaves; when others were obtained with subdivisions of the lower lobes in two or three, thus showing the palmate shape of *Aralia* leaves. the confidence in the value of the characters at first recognized had to be abandoned.

The first exposition of the Dakota Group flora shows four species of Ferns, six species of Conifers, and one of *Cycadeæ* only. To this small number we have added in this volume one species of *Gleichenia*, six species of Conifers, and five of *Cycadeæ*. The specific values of some of the vegetable remains referable to the Conifers is, however. doubtful, especially for those which are represented by cones only. *Abietites Ernestinæ*, *Sequoia formosa*, *Sequoia Reichenbachi*, and the fragments described as *Inolepis* are of this kind; all, however, though their specific or generic relation may be uncertain, are evidently representatives of some species of

Conifer. The fragments referable to this group are difficult of determination, for the organs represented upon the coarse shale or hard ferruginous sandstone of the formation merely expose some traces of their more prominent outlines, originally printed upon the soft embedding matter. We do not find, therefore, any flattened cones with the scales, nor any flattened branches with leaves, but impressions only, more or less deeply carved into the stone, the cones even passing vertically or obliquely through the shales and showing the space originally occupied, as a mere cylindrical hollow, around which the forms of the scales are more or less clearly molded. The numerous leaves of *Pinus* spread upon the surface have dug in the same way, and by their hard substance, narrow linear channels, representing the back of these leaves, with an indistinct midrib; and branchlets of *Sequoia* also are seen as longitudinal grooves, bearing on both sides the same impressed form of their leaves. This cannot be considered a very distinct representation of characters, the minute details desirable for an exact determination being more or less obsolete.

Among the specimens recently examined, a second fragment has been found referable to *Phyllocladus*.[1] The presence of this genus in the Cretaceous flora is thus sufficiently ascertained. We may, therefore, record as recognized in the flora of the Dakota Group, for the Ferns, the genera *Lygodium*, *Sphenopteris*, *Hymenophyllum*, and *Gleichenia*, the first three by each one species, the last by two; in the *Cycadeæ*, *Podozamites* by six species, and in the Conifers, *Sequoia* by three species, *Pinus* by one, *Phyllocladus* by one, *Torreya* and *Thuites* each by one, leaving out as of uncertain generic relation with the cones mentioned above, *Glyptostrobus* (?) *gracillimus*, which is perhaps identifiable with *Sequoia condita*, or with *Frenelites*, and *Geinitzia* (?), known merely by the impressions of some detached scales. To this should be added *Araucaria* from a species described in "Extinct Floras of North America" by Dr. Newberry, from Nebraska specimens.

The first dicotyledonous leaves described in the "Cret. Fl.," under the name of *Liquidambar integrifolium*, have been considered by some

---

[1] Since this was written, Heer, in part 2d of Vol. VI of the "Arctic Flora," has described this species under the name of *Thinfeldia Lesquereuxiana*, as a plant of uncertain relation.

authors as uncertain in regard to their generic relation merely on account of their entire borders. The form of the leaves, however, especially as figured (pl. xiv, fig. 3), with the lobes slightly enlarged above the sinuses, then gradually narrowed to a slightly obtuse point, and the nervation also, have the same character as those of the living *Liquidambar Styraciflua*. It is true that the four species of this genus known in the present flora have serrate borders of leaves. But three fossil species represented by leaves with entire borders have been described as *Liquidambar* from the Tertiary of Europe; and, though this reference is more or less hypothetical and controverted, it shows, nevertheless, that botanists of high standing— Unger, Watelet, Massalongo—have considered it, at least, as probable. It is easily seen that the leaves of *Aralia Towneri* (pl. vi, fig. 14) have a relation in shape or general outline to those of *Liquidambar integrifolium*, and this apparent similarity can but suggest the possible relation of all these and like forms to the genus *Aralia*. I may admit this relation as probable for the two leaves figured in "Cret. Fl.," pl. xxix, figs. 8 and 9, which are comparable, by their primary nervation, to those of *Aralia concreta* (pl. ix, figs. 3, 5). But though we have now a large number of specimens referable to diverse Araliaceous types, there is none as yet with leaves divided into lanceolate acute lobes like those which are figured in pl. ii, "Cret. Fl.," and with five primary nerves from the base. The reference of these leaves to *Sterculia* has been proposed also, from analogy of forms. But according to the definition of this genus as I admit it for the fossil leaves of the Dakota Group, I refer to it merely tripartite leaves with narrow linear lobes, comparable to those of *Sterculia labrusca*, like those of the few species described in this volume.

A number of vegetable remains of the Cretaceous are evidently referable by their characters to *Populus*. The only fragments of dicotyledonous leaves recognized by Heer, in the specimens which he studied from the Lower Cretaceous formations of Greenland (Kome), represent a *Populus*, appropriately specified by the name of *P. primœva*. From a higher stage of the same Cretaceous formation of that country (Atane) the celebrated Swiss paleontologist has described four other species of *Populus*. In his "Phyllites Crétacées du Nebraska," and from specimens of the Dakota Group, he has recognized *Populus litigiosa*, *Populus* (?) *Debeyana*, and another species still,

P. cyclophylla, described in Proc. Acad. Nat. Sci., Philadelphia. Professor Newberry, in his paper "On the Later Extinct Floras of North America," has described, also, besides the doubtful *P.* (?) *Debeyana*, three new species: *Populus* (?) *cordifolia*, *P. elliptica*, and *P. microphylla*. The specification and the interrogative punctuation applied to some of these names show that the authors themselves do not consider the generic reference as definitive, the character of some of the leaves being somewhat in disagreement with those generally recognized in species of *Populus* of our present time. Indeed, species of this kind, like the present *P. alba*, for example. have such multiplied and diversified forms of leaves, such great variability in their nervation, the mode of attachment, the length of the petiole, etc.. that they readily offer, by comparison with fossil leaves of obscure relationship, some points of affinity which, not being found elsewhere, have to be considered by the authors. Hence the doubtful references which may be. and are often, rectified by subsequent discoveries, as is proved by the great proportion of synonyms appended to the enumeration of *Populus* species. To obviate this inconvenient multiplication of fluctuating species of *Populus* I proposed a new generic division, under the name of *Populites*. for the classification of those Cretaceous leaves, numerous indeed, which, partaking of some of the characters of *Populus*, are nevertheless removed from this division by some others, as remarked in the first memoir which I published on some Cretaceous plants from Nebraska.

This paper had to be prepared on short notice from a limited number of specimens. but since its publication I have had opportunity to study the specific forms of the Cretaceous Flora by comparing a very large number of specimens, and have thus been able to recognize a more evident affinity of some of those leaves referred to *Populites* with other generic divisions. *Populites Lancastriensis*, *P. elegans*, which Schimper admits as a true *Populus*, and *Populites cyclophyllus* are the only species preserved in this genus. *P. oratus*, considered as possibly referable to *Celtis* in Cretaceous Flora. being rather related by its characters to the *Ampelideæ*, is described under the new generic division of *Ampelophyllum*. The affinity of *P. quadrangularis* being more evident with *Alnus*, has been described as *Alnites*. *P. flabellata*, as seen from other specimens, appears to be a deformed leaf of *Greviopsis Haydenii*, and *P. Salisburiæfolia*, being related to *Cissus*. is described as *Cissites*.

In regard to the distribution of *Populus*, to which are referred the most ancient dicotyledonous leaves known as yet, from the Lower Cretaceous of Greenland, the genus has, as said above, three species known already in the Upper Cretaceous of that same country, and five or six in the Dakota Group. It has, however, not been remarked in any Cretaceous Flora of Europe. It is not mentioned in the review of the genera represented by the, as yet, undescribed species of Aix-la-Chapelle,[1] and no form even distantly related is described in the Lower Paleocene Flora of Gelinden. It has, however, one species in the Eocene Flora of Sézanne, and increases in the number of its representatives in all the stages of the European Miocene. As far as we know it, till now, it has few species in our Lower or first American Tertiary Group—the Eocene; has a large proportion, eight per cent. of the species, in the Evanston Group; still more, or twelve per cent., in the Miocene of Carbon, and is present in the second, the Green River Group in four species, three of them of peculiar types, one of which is very abundant.

The presence of Willows (*Salix*) in the Flora of the Dakota Group cannot be controverted, though neither seeds nor scales of this genus have been found as yet. As it is seen in "Cret. Fl.," p. 60, pl. v. figs. 1–4, I have described as referable to one species only a number of leaves somewhat different in size and shape. As the specimens representing them are from the same locality, and as I recognized upon some numerous fragments of leaves a unity of character, size, form, and even texture and color, I considered them as mere varieties of leaves of the same tree. Dr. Newberry has, from the same formation, four species which, he says, he has chosen to regard as distinct, for geological convenience. No *Salix* has been recognized as yet in any stage of the Cretaceous of Greenland: but one species, *Salicites Hartigii*, Dkr., is from the Quader-sandstein of Germany, and another, *Salix Gœtziana*, Heer, from Quedlinburg. The genus is therefore sparingly represented in Europe and North America in Cretaceous Floras which are considered as nearly synchronous.

The other genera of the *Amentaceæ*, *Betula*, *Alnus* or *Alnites*, *Myrica*,

---

Quercus, Fagus, and *Ficus,* to which leaves have been referred in the Cretaceous Flora, do not require any observations. In this case, as in all the determinations of fossil plants, the characters of the species are not always satisfactorily established, but the generic affinities have been recognized or passed by authors without any marked criticism. The generic relation is specially positive for the remains referable to *Myrica;* one fragmentary leaf and some seeds have been already described in the "Cret. Fl.," while two fine new species are added in this memoir. It seems equally so for *Quercus* or its peculiar division, *Dryophyllum,* of which we have two new species, and for *Ficus,* to which three species are added.

Specimens of leaves referable to *Platanus* have been found in moderate proportion both in Nebraska and Kansas. The first was described by Heer, in the "Phyllites Crétacées du Nebraska," as *Platanus Newberryi,* from a very incomplete fragment. The accuracy of this determination was, however, subsequently verified by the discovery of more complete leaves, figured in "Cret. Fl.," pl. viii, figs. 2 and 3, and pl. ix. fig. 3, which show the narrowed base descending along the petiole lower than the point of union of lateral primary veins, and also the tendency to a three-lobed division, characters which are not observable in the fragment which Professor Heer had for his examination. To this fine species have been added: *Platanus primæva,* described from leaves so remarkably similar to those of *P. aceroides* of the Miocene that I was at first disposed to consider them as identical. I have lately received numerous large leaves of this species with specimens bearing fruits, which, very small, show a great difference from those of the living species; then, *P. Heerii,* rare, like the former, and found, as yet, only along the bluffs of the Salina River; *P. obtusiloba,* from a number of somewhat fragmentary specimens from Beatrice, Nebraska, representing leaves of about the same size and of the same characters; and *P. diminutiva*—all species described and figured in "Cret. Fl." The last one, as remarked in its description, may be a dwarfed form of *P. primæva* or *P. Heerii.* The leaf appears as gnawed along the veins by insects or perhaps by a parasite fungus. Its specification is not positive and is subject to criticisms. The base of the leaf is rounded to the petiole, a character as yet unique for a species of this kind. *P. recurvata* should, following the opinion of my honored friend Saporta, be referred to the *Araliaceæ* by a more intimate affinity to *Araliopsis* species;[1] and *Platanus affinis* seems now, after the examination and comparison of a number of specimens from Kansas, more evidently referable to the *Ampelideæ* than to the *Plataneæ*. Therefore these last two species are now eliminated from this generic division. The first is now *Araliopsis recurvatus*, the second *Cissites affinis*.

I persist in considering *Platanus Heerii* and *P. obtusiloba* as two different species, though it has been suggested that the last was probably a mere variety of the first. The identity is denied not only by the size, the *facies*, and the nervation of the leaves, but especially by the thinner texture of those of *P. obtusiloba*. The fact that the numerous specimens representing it are all from the same place in Nebraska, and that *P. Heerii* has not been found in that State thus far, confirms this separation. In regard to this last species Professor Geinitz has remarked in "Isis," 1875. p. 558, that paleontologists might, perhaps, recognize in it a *Credneria*. There is, indeed, some similarity in the general outline of the leaves. But this might be said of many of the generic forms of the Cretaceous, which seem referable to a few different types, or to present in one leaf the characters which are now generally found isolated in separate vegetable groups. The genus *Credneria*, known as it is to me by what is described by Stiehler. Vol. V of the "Paleontographica," includes species with cordate or sub-cordate leaves (none narrowed to the petiole), and bearing above the base two or three true secondary veins at right angles to the midrib. In *P. Heerii* the leaves are cuneate at the base, even gradually narrowed or decurrent on the petiole, which thus becomes slightly winged, and the veins under the primary nerves are mere marginal veinlets. Perhaps the relation of this species is more marked to the genus *Ettingshausenia*, which. I regret to say, is known to me only by supposed synonyms *Chondrophyllum grandidentatum*, as represented by Heer in the Cretaceous Flora of Moletein. and by *Phyllites repandus*, Sternb., two forms which have no affinity to *Platanus*.

The typical character of the Cretaceous species of *Platanus* is more evidently related to the Araliæ than to any other. This is proved by the reference to that genus of leaves now generally admitted as species of

---

[1] HEER, in "Arctic Flora," vol. vi, part 2, admits it as Sassafras.

Aralia, as *Platanus grandifolia, P. digitata, P. Jatropœfolia, P. Hercules,* Ung., and *P. latiloba,* Newby. The leaf of *Sassafras (Araliopsis) Platanoides* (pl. vii, fig. 1) has the *facies* and some of the characters of *Platanus* more distinctly defined than any other of the group; the same characters are even reproduced in *Aspidiophyllum platanifolium* (pl. ii, fig. 4).

The geological distribution of the genus *Platanus* is truly remarkable. No trace of it is recorded as yet in the Cretaceous of Europe, not even in the Paleocene and Eocene of France, so rich in fossil vegetable remains. Its first appearance in Europe is in the Upper Miocene of Oeningen, and of Austria and Italy, where it is represented by two very similar forms, *Platanus Guillelmæ* and *P. aceroides,* two species present in the same formation from the northern parts of the arctic lands to Italy. It is followed in the Upper Tertiary, or Pliocene, of this last country by *Platanus Academiæ,* Gaud., related as progenitor, perhaps, to the living *P. orientalis.* I have remarked above that the relation of leaves of the Dakota Group to *Platanus* has been considered as doubtful by some European paleontologists. This doubt may have been induced by the understanding of the total absence of *Platanus* leaves in the Cretaceous and Lower Tertiary of Europe. If so, it is certainly removed by the presence in our lignitic Eocene of some very beautiful and well characterized species of this genus: *Platanus Haydenii* and *P. Reynoldsii,* Newby. These species, discovered first in the Tertiary of the Upper Missouri River, near Fort Union, are predominant at Golden, Colorado, and are also found at Black Butte Station. The third Tertiary Group, that of Carbon, has, for the more numerous representatives of its Flora, leaves of *Platanus aceroides* and *P. Guillelmæ.* No species of this genus has been described from the Oligocene Green River Group; but we have from the Upper Tertiary (*Pliocene*) of California very fine specimens of leaves of two species, *P. appendiculata* and *P. dissecta,* closely related by their characters to the living *P. occidentalis.* Therefore, and considering the geological records, we may trace the origin of *Platanus* as far down as the North American Cretaceous, and follow its development through nearly all the stages of its Tertiary to our present time, by a number of closely allied intermediate forms.[1]

---

[1] *Platanus Heerii,* L. and *P. affinis* L. are mentioned by Heer in the Cretaceous of Atane, Greenland.

Coming now to the *Laurineæ*, I have to remark somewhat more definitely on the Cretaceous species referred to this family. The relation of some of them to the genera to which they have been referred is generally acknowledged, and the presence of the *Laurineæ* in our Cretaceous Flora receives a kind of historical authority from that of a *Sassafras* in a Cretaceous formation of Greenland,[1] of three species of *Daphnophyllum* in that of Moletein, and of *Laurus cretacea, Daphnogene primigenia, Daphnites Göpperti*, in that of Niedershoena. Of the species which have formerly been described in the Flora of the Dakota Group, *Laurus Nebrascensis* is related to *Daphnophyllum ellipticum* and *D. crassinervium* of Heer, while *Cinnamomum* and *Oreodaphne cretacea* are comparable to *Daphnogene primigenia* of Ettingshausen. *Persea Sternbergii* is also evidently of the same family, and the two leaves, described here below under the name of *Laurus proteæfolia*, are, indeed, allied to species of *Laurus* or of *Persea* by their nervation, especially by the more acute angle of divergence of the lower veins, though they show in the grooved middle nerve a character often remarked in species of *Ficus*, especially *Ficus protogæa*, Heer, of the Greenland Cretaceous Flora. Moreover, the fruit described ("Cret. Fl.," p. 74) as *Laurus macrocarpa* satisfactorily completes the evidence afforded by the leaves of the existence of species of *Laurineæ* in the vegetable world of the Cretaceous epoch. We have, however, to eliminate from this family *Laurophyllum reticulatum*, which appears more properly referable to *Ficus*. Its nervation, and especially its areolation, formed of square or irregularly polygonal meshes by the interposition of tertiary veins between the secondary ones and parallel to them, and the rectangular subdivision of its branches, are of the same character as in *Ficus Geinitzi*, Ett., *Ficus protogæa*, Heer, and as in many species of this genus now growing in Cuba, and even Florida, *Ficus suffocans, F. lentiginosa, F. pertusa, F. dimidiata*, etc. Numerous specimens recently found in Kansas represent the fossil species in characters more precise than formerly, as seen in its more detailed description under the name of *Ficus laurophyllum*.

But if the reference of some of the above-mentioned leaves to the *Laurineæ* is not contested, it is not the same in regard to those which, at

---

[1] In "Arct. Fl.," vol. vi, 2d part, pp. 75-78, HEER describes as new species *Laurus plutonia, L. angusta, L. Hollæ, L. Odini*, with *Cinnamomum Sezannense*, Wat., from the Upper Strata of Atane.

first appearance, were considered as more positively related to this family, and which have been described under the generic name of *Sassafras*. The question of the relation of those leaves which, by their number, seem to be the essential components of the North American Cretaceous Flora, has been already touched upon ("Cret. Fl.," p. 77). But since the publication of that work I have obtained from divers localities a large number of specimens of all the forms described there as species, and I have now some more data to offer to the consideration of paleontologists on the subject.

From historical documents the presence of *Sassafras* species in the Flora of the Dakota Group is as legitimately presumable as that of species of *Laurus* or *Persea*. In his "Flora fossilis arctica," Heer has described as *Sassafras arcticum* a leaf which, by its form, is similar to those described as *Sassafras cretaceum*, as remarked by the author, differing merely by its base tapering somewhat less narrowly to the petiole. The nervation is of the same character. Saporta considers the Greenland leaf as a true representative of *Sassafras*. He has himself published in the "Sézanne Flora,"[1] as *S. primigenium*, two fragmentary leaves whose base, more narrowly tapering, is similar to that of *S. Mudgei* of the "Cret. Fl.," as well as the lobes which, enlarged in the middle, have that ovate-lanceolate shape so distinctly marked in the present *S. officinale*. There is also no appreciable difference in the nervation. The lower secondary veins of the middle lobe ascend a little higher in the leaves of the Sézanne Flora, and unite with those of the lateral lobes somewhat nearer the borders of the sinuses. But in some of the specimens of Kansas the same appearance is remarked also, and the difference between the greater or less distance which separates from the sinuses the branches which unite the upper division of the secondary veins is observable upon leaves of *S. officinale*, this division being sometimes marginal, sometimes curving one to three millimeters lower than the border of the sinuses. Comparing leaves of *Sassafras officinale* with those represented by Saporta in the "Flora of Sézanne" and the specimens of *S. Mudgei* from Kansas, it is impossible for me to recognize any character, even any specific difference, by which these leaves could be separated. It is therefore not surprising that Dr. Newberry first, and after him Heer and Schimper, did consider Cretaceous

---

[1] P. 366, tab. viii, figs. 9 and 10.

specimens of this kind as representing species of *Sassafras*. In the last volume of his superb work on Vegetable Paleontology,[1] Prof. W. P. Schimper, speaking of leaves of *Sassafras cretaceum*, of which I had sent him photographical designs, remarks: "That those leaves, very variable in size, present such a remarkable likeness to those of *S. officinale*, now living in North America, that one would be disposed to consider them as belonging to a homologous species." He rightly adds that the only difference seems to be in the thicker substance of the fossil leaves. Even on this point I have from Texas specimens of the present *S. officinale*, whose leaves appear of a consistence nearly as thick as it seems to be in those of the Dakota Group.

On the other hand, no species of the *Laurineæ* family living at our time is known with dentate leaves; and it may be remarked, from the figures, that the two leaves described as *Sassafras cretaceum* ("Cret. Fl.," pl. xi, figs. 1 and 2) have the borders of the lobes somewhat dentate, and some of the secondary veins running into the point of the teeth, or craspedodrome. This character is still more marked in *S. mirabile, loc. cit.*, pl. xii, fig. 1, a form extremely common in Southern Kansas, and represented in very numerous and remarkable varieties. In some of the leaves the secondary veins are all camptodrome, and therefore the borders of the lobes are entire. In others, as seen, pl. xi, fig. 2, the outside lateral veins are craspedodrome, and thus the borders are dentate, while on the inside they curve along the borders, which are entire. In the fine complete leaf (fig. 1 of the same plate) the middle lobe has the veins all camptodrome on the left side, while on the right one, a few of them, one or two, reach to the border, which has, therefore, one or two short indistinct teeth, and the lateral lobes are clearly dentate on the outside only. This evidently shows such a disposition to variations of nervation and border divisions, that I formerly considered as unjustifiable a specific, and still more a generic, division between the leaves of pl. xi, figs. 1 and 2, and those of pl. xii, figs. 2 and 3, of the "Cret. Flora." When, therefore, we find the same difference between the leaves which represent *S. mirabile* (pl. xii, fig. 1), it seems that the same conclusion should follow. But in this case, with the more generally predominant character of the indentation of the leaves,

---

[1] Traité de Paléontologie végétale, vol. iii, p. 292.

which, in some specimens larger than the one figured, are now deeply cut by divisions like pointed lobes, there is still another character, remarked on specimens recently discovered, which seems more forcibly to separate these forms from the *Laurineæ*, and indicates a more evident relation to the *Araliaceæ*. A number of those specimens communicated by M. Chs. Sternberg, to whose careful and zealous researches the Flora of the Dakota Group is indebted for many important discoveries, represent large leaves, which, by the outlines, the nervation, and the dentate borders of the lobes, are like *S. mirabile* of pl. xii, fig. 1. The leaves, however, which are much larger, the lobes measuring as much as ten centimeters in length from the point of union of the primary nerves, greatly differ by the forking of the lateral nerves from a point two and one-half centimeters above their base, thus forming, of course, a subdivision of these lobes into two equal parts, or a palmately five-lobed leaf. They are described as *Sassafras (Araliopsis) dissectum*. Among the innumerable varieties in the shape of the leaves of the living *Sassafras officinale* we see a constant and gradual mode of division, passing from a round or oval and entire shape to a bilobed and trilobed one; but, as yet, I have been unable to observe a single case of subdivision of the lateral lobes, or to find a palmately five-lobed Sassafras leaf. This character is, on the contrary, far more generally seen in the *Araliaceæ* of our time. Even in a section of the *Araliaceæ*, the genus *Hedera*, whose leaves may be compared to some of those under examination, I do not know any species with trilobate leaves. *Hedera turbascens, H. discolor, H. argentea, H. aurifolia, H. jatropæfolia.* have leaves five to seven palmately lobed, or when occasionally trifid their segments are narrow and acuminate. From this the relation of the five palmate leaves to the *Araliaceæ* becomes more evident.

Going further into this kind of investigation, we are met by a new difficulty in the appearance of another modification in the character of this peculiar type of leaves. In examining the first specimens of the species represented (pl. xii and xiv), I could but consider them as representing either *Sassafras (Araliopsis) obtusum* or *S. mirabile*, the specimens being fragmentary, having only the lobes or part of them preserved. As long as the auricled and peltate base was unknown, the reference of the specimens could not be different. The nervation, the form of the lobes, their size, all are of the same character as in *S. mirabile.* But in the peltate base of the leaves there is another character which, separately considered, relates the leaves to the *Menispermaceæ.* We thus have *Sassafras* already represented in those leaves by *S. Mudgei,* and less positively by *S. acutilobum: Araliopsis,* to which are referable *S. mirabile,* with the dentate *S. cretaceum, S. obtusum, S. dissectum, S. platanoides, Platanus recurvata,* and in a new generic division, under the name of *Aspidiophyllum,* the leaves which, either *Aralia* or *Sassafras,* by their upper trilobate part, are necessarily separated from these genera by their auricled peltate appendage. Still, the subdivisions in the classification of the peculiar and so-called *Sassafras* leaves have to be pursued further, for by degrees and by the gradual obliteration of their lobes they become round or truncate, or broadly pointed at the top, preserving more or less the narrowed base, tapering to a long petiole, and the trifid craspedodrome nervation from a distance above the borders, and thus they become more evidently related to other vegetable orders. One species is a true *Hedera,* another passes to the *Hamamelideæ,* and a number have their affinity with the *Ampelideæ.*

The characters of the leaves of the *Ampelideæ,* especially those of *Cissus,* are somewhat obscurely represented in *Sassafras Harkerianum* ("Cret. Fl.," pl. xi, figs. 3 and 4; pl. xxvii, fig. 2) and in *S. obtusum* (pl. xiii), more distinctly in *Cissites acuminatus* (pl. v, fig. 3) and *C. Heerii* (pl. v, fig. 2), two new species described in this memoir. They appear to constitute an indivisible group. Some of the leaves formerly described as *Populites* are also referable to this section, or to another less exactly defined *Ampelophyllum,* allied by some of its characters to *Hedera,* by others to *Credneria,* thus intermediate between the *Ampelidæ* and the *Tiliaceæ;* by the areolation this genus is related to *Greviopsis,* and also more distantly to *Chondrophyllum* of Heer, as remarked in the description. From this it is perceivable that this *Sassafras* type, which at the beginning was regarded as simple, well defined, and limited in its character, is, on the contrary, multiple, and representing forms which, from increased researches and discoveries, indicate affinity to a number of different genera or orders of the vegetable kingdom.

The same remark is equally applicable to the leaves which have been described in the "Cret. Fl." under the generic name of *Protophyllum.* The

c r 2

disagreement in the affinities of its species has been explained in the
remarks following the description of the genus. I have now to add still
to this division two leaves recently communicated from Kansas, represented
in pl. iii, fig. 1, and pl. viii, fig. 4. They fully confirm the former observations. By the outline of the leaves, their craspedodrome nervation,
and the presence of two pairs of secondary veins under the primary ones
and at a right angle to the midrib, they represent a species of *Protophyllum;*
but the border base of the leaves is truncate, not subpeltate, and by this
difference the leaves are rather referable to *Credneria,* from which, however, they differ by the veins as well as their divisions, being all craspedodrome, and by the truncate, not cordate, base of the leaves. I formerly
published a short description of them under the name of *Credneria? microphylla.* It now seems that, by their evident relation to *Protophyllum quadratum,* they have to be admitted in this last generic division, an opinion
which may be put at naught by the discovery of specimens pointing to
another reference for these leaves.

We have, also, an addition of three new species to the group of Cretaceous plants described under the generic name of *Menispermites.* In
this case, however, there is no difficulty whatever in conformably uniting
into a definite group the characters of the leaves which, round, ovate, or
oval, with borders entire or undulate, have a common generic affinity,
indicated by their nervation. In order more clearly to bring into view the
relation of the undulate-lobed forms of leaves described in the "Cret. Fl."
(pl. xx, figs. 1–4, and pl. xxv, fig. 1), I have represented (pl. xv, fig. 4)
a finely and wholly preserved leaf of *Menispermites obtusilobus,* which,
though small, is easily identified with the large one of "Cret. Fl." (pl. xxv,
fig. 1). Now, comparing it to figs. 2 and 3 of the present pl. xv, the
identity of nervation is defined by the five basilar veins, with a thin pair
of marginal veinlets underneath; and by the upward direction of the
internal lateral veins, which in fig. 4 ascend to above the middle, pass
still higher in the short oval leaf, fig. 3, and reach nearly to the obtuse
point in fig. 2. The subdivision of the tertiary veins is in all the leaves
of the same type, and the shape of the leaves or their outlines are mere
modifications, depending upon the direction of the veins. The leaf, fig. 3,
is peltate from the point of attachment of the petiole near the middle.

The character of the nervation remains, however, the same. It is somewhat obscured in the figure from indistinctness of the specimen. In figs. 1 and 2, representing leaves entirely preserved and nearly round, the nervation is marked by three pairs of primary nerves on each side of the midrib, and under them by one pair of true marginal veinlets curving on each side toward the borders. Comparing, therefore, these peltate leaves with fig. 4, the position of the petiole is the only notable difference, and the transition to fig. 5 by slight modifications of characters is easily remarked. The peltate form of these round leaves has suggested the fitness of a slight modification in the characters assigned to the genus *Pterospermites* in the "Cret. Fl." (p. 94), the leaves being sometimes rounded or subcordate at base. The difference is immaterial, and is remarked even upon leaves of the same species of *Menispermum* of our epoch. These round peltate leaves, for example, are so much like those of living species of *Cissampelos*, that they rather prove the adaptation of this generic division to all the Cretaceous leaves which I have referred to it.

The *Magnoliaceæ* are more numerously and definitely represented in the North American Cretaceous Flora than they are in that of Europe. *Magnolia alternans* and *M. Capellini* have been described by Heer in his "Phillites Crétacées du Nebraska;" and since that time these two species have been recognized throughout the whole explored area of the Dakota Group, as also in the lower stage of the Cretaceous of New Jersey, and in the Upper Cretaceous of Greenland. *M. speciosa* of Moletein has been discovered in Colorado with a fruiting cone or carpite of this genus. Two other species have been described from the Dakota Group: one, *M. obovata*, by Dr. Newberry, in his "Ancient Floras," another, *M. tenuifolia*, in "Cret. Fl.," and two new ones, *M. obtusata* and *M. Isbergiana*, by Heer, from Atane. In Europe, *M. amplifolia* and *M. speciosa* are described by Heer in the Flora of Moletin—there represented by leaves and fruit.

To the same order belongs *Liriodendron*, so easily recognized by the peculiar form of its leaves. Its Cretaceous origin, or rather existence, is marked in the Dakota Group by a number of specific representatives locally and distantly distributed. The genus is not represented in the Cretaceous Flora of Europe; but in the "Cretaceous Flora of Groenland" Heer describes six varieties of *Liriodendron Meekii* from Atane, and no less than eight specific forms have been described from Nebraska and Kansas—some of them extremely well defined. This shows, perhaps, more evidently than any other fact remarked on the characters of the plants of the Dakota Group the great disposition to variableness by modification of some characters in the first Dicotyledonous plants. These changes have either caused a multiplication of specific forms preserving traces of the original types in traversing the subsequent geological formations, or have gradually destroyed the number of specific representatives of some genera, as in *Liriodendron*, or even caused the total disappearance of some of the best defined and more predominant types, like those of *Credneria, Pterophyllum*, etc. Of these, however, the original characters may have been so widely varied that the ultimate derived forms have not yet been distinctly recognized on plants living now. The two last-named genera, *Credneria* and *Protophyllum*, may possibly be referable to some subdivisions of the *Columniferæ*, the *Buttneriaceæ* and *Pterospermæ*, for example.

The three species which I have described under the insufficiently-defined genus of *Sterculia* are all very uncertain in their relation. As much may be said for the following and last classes of the vegetable kingdom:

To the *Acereæ* is referable *Negundoides acutifolius*. The leaf, however, as seen from pl. xxi, fig. 5, and its description, is too fragmentary for a satisfactory determination of its characters. *Acer antiquum* is described by Ettingshausen in his "Flora of Niedershœna," but from the opinion of the author the reference is uncertain. The leaf rather resembles a deformed form of *Quercus* or of *Liriodendron*. In the same order Heer has, from the Upper Cretaceous of Greenland, a *Sapindus prodromus*, represented by one leaf only, which has evidently the character of the genus. A beautiful species of *Sapindus* described here from Colorado is also present at Atane. This genus is therefore Cretaceous. The reference to the *Rhamnaceæ* of the leaf described as *Rhamnus tenax* in "Cret. Fl." is apparently legitimate, for of the same group three other species, *R. prunifolius*, a *Celastraphyllum*, and an *Ilex*, are described here from the same formation.

To the *Anacardiaceæ* we have probably to refer, as *Rhus Debeyana*, the species described as *Populus* and as *Juglans Debeyana* as seen in "Cret. Fl., p. 110. I have not obtained from the Dakota Group any new materials comparable to this form, especially common in Nebraska; but I have seen a very fine specimen of it got out of a deep tunnel in Oregon, presenting upon its surface small punctiform protuberances, apparently oily glands, like those remarked upon leaves of the living *Rhus aromatica* and other species of this genus. The leaves are figured (pl. lvi, figs. 5, 6). A species of *Rhus* is described from the Cretaceous of Greenland by Heer, while considering historical authority, we have the same evidence in favor of *Juglans* by a species of this genus in the Cretaceous Flora of Moletein and one in that of Greenland.

Of the *Rosifloreœ* we have from the Dakota Group one leaf and one fruit described as *Prunus*. I have recently received from M. Towner a fruit of the same character upon a specimen bearing leaves of *Aralia Towneri*.

The *Myrtiflorœ*, as well as the *Leguminosœ*, present by a number of specimens in the Greenland Cretaceous, have not been thus far positively recognized in Kansas and Nebraska, but seen by one silique only in Colorado.

The few groups not considered in this review have been remarked upon already in the "Cretaceous Flora," and the views in regard to the leaves referred to them have not been modified either by remarks of European authors or by the discovery of new materials.

The want of positiveness in the characters of some of the Cretaceous plants cannot in any way weaken reliance upon the data derived from the exposition of the Flora of the Cretaceous age, nor throw any discredit on the conclusions which they dictate. What the Flora of the Dakota Group positively shows is a great predominance of dicotyledonous plants in its composition; and that is all that may be positively known as yet of the remarkable change it attests in the vegetation of that period. The causes, the mode of proceeding of nature, either by slow, gradual, or by rapid modifications, remains as yet inscrutable. But the characters of dicotyledonous leaves cannot be mistaken; the relation of most of them to groups of plants of the present Flora possesses positive evidence. The *Cupulifereœ* with species of *Quercus* and *Fagus;* the *Salicineœ* with species of *Populus;* the *Plataneœ* with *Platanus primœva*, leaves and fruits; the *Laurineœ*, represented also by leaves and a fruit of *Laurus*, by leaves of *Persea, Cinnamomum, Sassafras;* the *Araliaceœ*, the *Magnoliaceœ*, with fruits and leaves;

the numerous forms of leaves of *Liriodendron*, so peculiar that they cannot be mistaken for those of any other group or plant; even the *Menispermaceæ* constitute, by their fossil remains, vegetable groups quite as definite as they could be established from living plants.

Since the publication of the "Cretaceous Flora" (vol. vi of the U. S. Geological Reports of Dr. F. V. Hayden) the character of the vegetation of the Middle Cretaceous as represented in the Dakota Group has become better defined by the discovery of a large number of specimens of fossil plants, which have increased from 130 to 190 the number of vegetable forms considered specific, already known from this formation. The whole Flora of the Cenomanian epoch, as it is shown in the table of distribution, is composed of 446 species, of which 310 are dicotyledonous and 130 are cryptogamous and gymnospermous plants. Of the 190 species of the Dakota Group, 162 are dicotyledonous and only 28 represent crytogamous and gymnospermous plants.

Numerous works on the Jurassic Flora have sufficiently proven that up to its upper member the Wealden, or lower Neocomian, it is entirely composed of gymnospermous and cryptogamous plants—especially Ferns, *Cycadeæ*, and Conifers. The Neocomian, whose vegetation is but little known as yet, shows in its remains the same constituents of its Flora. Upon it is superposed in Germany the upper Neocomian, or Urgonian, from which a series of fossil plants, 22 in number, have been described by Schenk from the Wernsdorf-Schichten of the Carpathian Mountains of Austria; and there also no dicotyledonous plant has been found, and nothing indicates the decadence of the reign of the gymnospermous plants or shows any kind of difference which could lead one to presage the appearance of the Dicotyledons.

We owe to Heer the most interesting documents on the characters of the vegetation of the Middle Cretaceous—first by the publication of the Flora of Kome, and then of that of Atane, both in Greenland.

The Flora of Kome, composed of 85 species, has, says the author, its greatest affinity with that of the Wernsdorf shale or upper Neocomian on one side, and with that of the Wealden on the other. With the plants of the higher Cretaceous stages it has only 7 species—Ferns and Conifers—in common. Most of the specimens of the group submitted to Heer's examination have been obtained on the peninsula of Noursoak (70° 37' N.), from beds of shale alternating with banks of sandstone, the whole over-lying granite or primitive formation. One of the localities, that of Elkor-fat, is 500 feet above that of Kome, but the plants are of the same kind. The vegetable remains belong mostly to cryptogamous and gymnosper-mous plants: 41 Ferns, 1 Marsilia, 1 Lycopod, 3 *Equisetaceæ*, 10 *Cycadeæ*, 21 Conifers, 6 Monocotyledons, and a single Dicotyledonous species.

On the south side of the same peninsula of Noursoak, near Atane, at an elevation of 650 feet above the sea, another lot of plant-remains, col-lected also by the expeditions of Nordenskjöld, and submitted to Prof. Heer for examination, represents a Flora composed of far different elements. It has 170 species: 3 Fungi, 31 Ferns, 1 Marsilia, 1 Selaginella, 1 Equi-setum, 8 *Cycadeæ*, 27 Conifers, 8 Monocotyledonous, and 97 Dicotyledonous plants. These, therefore, constitute more than one-half of the vegetation.[1] The celebrated author remarks, on the geological relation indicated by the characters of the plants, that it is not possible to determine it positively, as the plants of the Cretaceous are, as yet, too little known. But he admits that the formation of Atane, considering its vegetable remains, is probably referable to the lower Cenomanian.

As will be seen in the examination made of the age of the Dakota Group, from data shown in the table of distribution, its Flora seems to be somewhat more recent than that of Atane, though the relationship is very close. The general character of the plants does not greatly differ, but the number of the dicotyledonous plants is much greater, amounting in the Flora of the Dakota Group to more than five-sixths of the vegetation.

In considering merely what is now known of the vegetation of the Middle Cretaceous (the Cenomanian of d'Orbigny), the first appearance, and especially the prodigious development, of the Dicotyledons seems the more wonderful that it is not a local phenomenon, but is remarked in the formations of the same age over the whole Northern hemisphere. We cannot yet follow it in all the intervening land areas, but it has been traced from Greenland to Vancouver Island to Canada, to Kansas, and Colorado, and in Europe to Germany, therefore in about 40° N. latitude.

---

[1] These data are taken from Heer's "Groenland Flora," vol. vi, part 2.

With the limited acquaintance we have with the ancient Floras of the world it is not possible to account for the sudden appearance of the Dicotyledons in the Cretaceous time and for their rapid and wide distribution. Saporta, justly considered as the botanist who has acquired by his vast knowledge the most extensive views on the distribution of the vegetation in the ancient epochs, says, on the subject:[1] "The organic evolution to which the Dicotyledons owe their existence and their distribution must have been produced under the influence of very different conditions. It is possible that the evolution has been originally slow and obscure; possibly also it has been accomplished in a concealed or as yet undiscovered locality, in a separate region, and under the influence of peculiar local circumstances. It is probable that the change may have been accomplished by the mediation of insects, multiplying at a given time the results of crossing and producing some combinations favorable to the growth of these plants. It is even conceivable that a short time may have been sufficient to give origin to plants of this class under the action of causes which are still unknown. Whatever hypothesis may be preferred, the fact of the rapid multiplication of the Dicotyledons and of their simultaneous occurrence in many localities of the Northern Hemisphere from the beginning of the Cretaceous Cenomanian cannot be contested."

Yes, in this case, as in many others, we may collect facts, but the work of nature in its mode of proceeding for the creation or modification of species remains inscrutable. We may consider the formation of the Dakota Group as produced by a very slow, gradual, prolonged depression of the Western slope of the continent, bringing up from the South or West the invasion of ocean water charged with muddy materials, periodically heaped farther and farther inland by powerful tides. We may suppose, too, the invading flow as bringing with it seeds or fragments of roots of plants derived from a country now covered by the sea, and distributing here and there those germs of vegetable organisms. But all this does not account for much in the solution of the problem; it may explain the distribution; but the first appearance, and it seems the simultaneous multiplication, of the dicotyledonous plants remains a fact inconceivable to reason.

---

[1] "Le monde des Plantes," etc., p. 197.

DESCRIPTION AND ENUMERATION OF SPECIES OF THE AMERICAN DAKOTA GROUP FORMATION.

I. CRYPTOGAMÆ.

THALLOPHYTES.

ZONARITES, Brgt.

Zonarites digitatus, Gein.

"U. S. Geol. Rep." vI, p. 44, pl. i, fig. 1.

The relation of this vegetable organism to that described by Brongmart and Geinitz is contested on account of the habitat in a different Geological stage, Geinitz having described his plant from the Dyas. As species of Thallophytes of the Devonian are represented by identical forms in more recent formations, even in the Cretaceous of Europe, the objection is not imperative.

ACROGENS.

EQUISETACEÆ.

EQUISETUM, Linn.

Equisetum nodosum, sp. nov.

Stems small, one-half to one centimeter in diameter, obscurely narrowly striate; articulations very inflated, marked with broad round scars of points of attachment of branches above the line of division.

The species is distantly related to *E. amissum*, Heer, "Fl. Arct.," III, p. 60, pl. xiii, figs. 2–8, of the Lower Cretaceous of Kome, essentially differing by the inflated articulations. It is represented by specimens Nos. 473, 536 of the Museum of Comp. Zool., Cambridge; they are too small and obscure for definite comparison.

Hab.—7 miles N. E. of Glasco, Kansas; collected by *Chs. Sternberg.*

FILICES.

SPHENOPTERIS, Brgt.

Sphenopteris corrugata, Newby.

"Later Ext. Fl. of North America," p. 10: "Illustr.," pl. ii. fig. 6.

Frond unknown; pinnules ovate or cuneiform, narrowed at the base, obtuse, lobed, and often plicate longitudinally; nerves distinct, dichotomous, branching from the base.—(Newby.)

HYMENOPHYLLUM, Klf.

Hymenophyllum cretaceum, Lesqx.

"U. S. Geol. Rep.," vi, p. 45, pl. i, figs. 3-4b; xxix, fig. 6.

In describing this species I related it to the preceding from the description given by the author, as I had not then seen the figures. These indicate a degree of relation which cannot be positively ascertained on account of the too fragmentary specimens. The fronds were evidently large in the plants of this kind. The divisions are multiple and extremely variable. The specimens may, therefore, represent pinnules derived from divers parts of fronds of the same species.

PECOPTERIS, Brgt.

Pecopteris Nebraskana, Heer.

"U. S. Geol. Rep.," vi, p. 46, pl. xxix, figs. 5, 5a.

GLEICHENIA, Sw.

Gleichenia Kurriana, Heer.

"U. S. Geol. Rep.," vi, p. 47, pl. i, figs. 5-5c.

Gleichenia Nordenskiöldi, Heer.

Plate I, Figs. 1, 1a.

Hayden's "Ann. Rep.," 1874, p. 334, pl. ii, fig. 5.

Fronds slender, bi-polypinnate; ultimate pinnæ alternate, rigid, open, linear, parallel; pinnules subcoriaceous, small, free, oblong-ovate, obtuse, rounded at base on both sides, inclined upward; secondary veins few, three or four pairs, the lower forking, the upper simple.

Though the American specimens of this species are small they show distinctly the essential characters of the species: the slender rachis of the ultimate pinnæ rendered flexuous by compression of the basilar border of the pinnules, the very small leaflets free and rounded at base, and the disposition of the veins. The specimens which I have for examination are sterile. As seen and figured by Prof. Heer, the fructifications are those of the subgenus *Didymosorus*, Deb. and Ett., two sori placed upon the middle of the lower pair of veins, one on each side of the medial nerve.

The rachis of this fern is described by Heer as slender. As it is figured here it appears somewhat broad, though not larger than it is represented in Heer's "Fl. Arct.," iii, pl. ix, fig. 6. The ultimate rachis is, however, very slender filiform.

Hab.—Fort Harker, Kansas. *Chs. Sternberg.*

LYGODIUM, Sw.

Lygodium trichomanoides, Lesqx.

"U. S. Geol. Rep.," vi, p. 45, pl. i, fig. 2.

PHŒNOGAMÆ.

GYMNOSPERMÆ.

ZAMIÆ.

PODOZÁMITES, Fr. Br.

Fronds pinnate; leaves distant, obliquely or horizontally attached by an attenuated pedicelliform half-twisted flat base articulated upon the rachis and therefore caducous; veins equal, longitudinal, converging to both ends of the leaves; borders entire.

This genus of Braun, as amended by Saporta and Schimper, seems adapted for the description of all the leaves of Cycadeæ found as yet in the Dakota Group.

Podozamites Haydenii, Lesqx.

Pterophyllum Haydenii, Lesqx., " U. S. Geol. Rep.," vi, p. 49, pl. i, figs. 6, 6*b*; Hayden's "Ann. Rep.," 1874, p. 334.

Nothing more definite is known of these vegetable fragments than has been published as quoted above.

Professor Heer, considering the thickness and impressions of the stems, regards these fragments as more probably referable to Conifers of the section of the *Araucarites* than to *Cycadeæ*. No leaves of this section, however, can be compared to those which I have figured, and which, by their parallel veins and forms, are very much like the leaves of some species of *Podozamites*. Indeed, from the remarks on this genus by Heer, the leaves are either narrowed and joined to the stems by decurring to it, or produced into a short pedicel attached to the stem by small tubercles or warts. The characters of the genus are thus exactly shown not only by the leaf, but also by the stem whose round small scars indicate points of attachment like tubercles.

The leaves closely resemble those of *Podozamites lanceolatus* as figured by Nathorst, "Fl. of Bjuf.," pl. xvi. fig. 3. I think, therefore, that the fragments figured in "U. S. Geol. Rep.," figs. 6 and 6*b*, should be referred to this genus. The relation of the cone, however, which I referred to the same species from its likeness to that of Stichler as *Pterophyllum Ernestinæ*. is wrong, as it evidently represents a Conifer.

Podozamites oblongus, Lesqx.

Plate I, Figs. 10, 11.

Leaves oblong, gradually narrowed from below the middle to the flat sessile base, rounded at the eroded apex; veins thin, parallel, close, equal, distinct with the glass.

These leaves are evidently overturned upon the plate. The apparently truncate lower part seems as an enlarged point of attachment similar to that of species of *Cordaites* of the coal. But the irregular erosion is accidental or caused by compression of the macerated apex of thick coriaceous leaves.

These leaves are of the same character as those of *P. lanceolatus*, Schp., in Heer, "Fl. Arct.," iv, pl. vii, figs. 1–7 of the Jurassic Flora of Spitzbergen, differing by the more abruptly rounded apex.

Hab.—Dakota Group, Kansas. *Chs. Sternberg.*

Podozamites angustifolius? Heer.

Leaves long and narrow, somewhat falcate or ensiform, linear-lanceolate, gradually slightly narrowed upward from the middle, obtuse? (point broken) more rapidly downward from below the middle to the point of attachment, distantly veined; primary veins obtuse, prominent; surface smooth, minutely lineate.

The preserved part of the leaves is 11 centimeters long, averaging 9 millimeters broad. The point of attachment is flat, 3 millimeters broad. As the apex of the leaf is broken it is not possible to see if it is acuminate.

The leaves figured by Heer, "Fl. Arct.," iv, pl. vii, figs. 8–11, and pl. viii, fig. 5, are either acuminate or somewhat obtuse and slightly scythe-shaped, as in the one described here, but this is broader than any of those of Heer. The nervation seems like that of *P. Eichwaldi*, Heer, *ibid.*. the primary veins being broad, thick, or prominent, so that the surface appears undulate and the intervals marked by irregular or not continuous very small veinlets. This leaf is also, from its shape and size, comparable to

Podozamites ensis, Nath., "Fl. of Bjuf.," pl. xv, fig. 2. This, however, has the veins narrower and indistinct.

Hab.—South of Fort Harker, 4 miles east of Minneapolis, and 7 miles northeast of Glasco, Kansas. *Chs. Sternberg.*

Podozamites prælongus, sp. nov.

Leaf large, oblong, linear, narrowed gradually to the point of attachment, obscurely veined; primary nerves parallel and distinct.

The upper part of the leaf broken at 12 centimeters from the base is there 5 centimeters broad. The whole length appears to be about 16 centimeters. Its size is greater than that of any of the leaves of this genus figured by authors, larger than the fragment of *P. lanceolatus-latifolius,* Heer, "Fl. Arct.," iv, pl. xxvi, fig. 6.

Though obtuse, the veins appear more distant and broader than in this last species.

Hab.—South of Fort Harker, with the preceding.

Podozamites emarginatus, sp. nov.

Leaves large, linear-oblong, gradually narrowed from below the middle to the flattened base, abruptly rounded and deeply emarginate at the apex; primary nerves parallel, distinct or prominent, conjoining at the apex and the base, separated by thin disconnected veinlets.

The leaf is 14 centimeters long, 3½ centimeters broad in the middle, the point of attachment 4 millimeters broad. It is abruptly rounded at the top to 2 centimeters broad and there deeply obtusely emarginate, the borders joining into a small obtuse sinus 1½ millimeters wide.

The emargination of the top may be a casual deformation, but even if the apex was regular and obtuse this species is without marked affinity to any other of the genus.

Hab.—Seven miles northeast of Glasco, Kansas.

Podozamites caudatus, sp. nov.

Leaf large, enlarged and oval in the middle, where it is 5 centimeters broad, rapidly narrowed to a point of attachment 7 millimeters broad, and attenuated from above the middle in rounding to a long acumen measuring 1½ centimeters broad at the point where it is broken 13 centimeters from the base.

The leaf has a peculiar form, being abruptly enlarged in the middle from above the base and as rapidly narrowed into a long linear acumen whose upper part is destroyed. The primary veins are flat and broad,

distinct, half a millimeter distant, with an indistinct veinlet in the narrow
intervals.

The form of this leaf is peculiar, without relation to any of this genus.
Hab.—Near Fort Harker, *Chs. Sternberg*. No. 117 of the National
Museum.

CONIFERÆ.

PHYLLOCLADUS, Rich.

Phyllocladus subintegrifolius, Lesqx.

"U. S. Geol. Rep.," vi, p. 54, pl. i, fig. 12; Hayden's "Ann. Rep.," 1874, p. 337, pl. 2, fig. 4.
Thinfeldia Lesquereuxiana, Heer, " Fl. Arct.," vi, p. ii, p. 37, pl. xliv, figs. 9, 10.

Figure 4 of "Ann. Rep." represents the lower half of a leaf of same
character as that in "U. S. Geol. Rep.," *l. c.*

ARAUCARIA, Juss.

Araucaria spathulata, Newby.

"Notes on Ext. Fl.," p. 3; "Illustr.," pl. ii, figs. 5, 5a.

Leaves close, broadly spathulate, obtuse, narrowed above the enlarged base,
carinate; medial nerve distinct at base, effaced from the middle upwards.—(Newby.)

The author remarks that the specimen represents a fragment of a
branch nearly half an inch in diameter on which the leaves are thickly
set, their base slightly decurring scarcely separated from each other.
From their base the leaves, half an inch in length, radiate in all directions.

The species is closely allied to *Abietites curvifolius*, Dkr., of the
Quader-sandstone of Blankenburg. This has the leaves rounded at the
apex, a deep medial nerve, and the leaf scars very distinct. This last
character is well marked on the figure of *A. spathulata*. The same figure
shows the leaves reflexed or spreading at the base, the only part seen.
In Dunker's species the leaves are curved up from the middle and are
longer.

Hab.—Sage Creek, Nebraska. Dr. *F. V. Hayden.*

TORREYA, Arn.

Torreya oblanceolata, sp. nov.

Plate I, Fig. 2.

Branches slender; leaves long, flat, gradually enlarging upwards from the decurring base; medial nerve thin.

The figure represents the best and largest of the fossil fragments.

None of them has a leaf entirely preserved, and thus the upper end of the leaf is undetermined.

From the decurring base of the leaves the fragment may represent a *Sequoia*. It has some analogy to *S. Smittiana*, Heer, "Fl. Arct.," iii, pl. xvii, figs. 3, 4, while *Torreya parvifolia* and *T. Dicksoniana*, Heer, *ibid.*, pl. xvii and xviii, have the leaves sessile, and in this last species rounded and enlarged above the point of attachment. The leaves of this fragment, however, are too long for a species of *Sequoia*, also flat, not rigid nor coriaceous, and thus seem referable to Torreya.

Hab.—Cretaceous black shale, near Golden. Rev. *A. Lakes.*

SEQUOIA, Endl.

Sequoia Reichenbachi, Heer.

Lesqx., "U. S. Geol. Rep.," vi, p. 51, pl. i, figs. 10, 10b.

The supposed relation of the cone referred to this species is contradicted by Professor Heer. Though the cone represents a *Sequoia*, the specific name is left undetermined.

Sequoia fastigiata? St.

Hayden's "Ann. Rep.," 1874, p. 335, pl. iii, figs. 2, 8, 8a.

Branches slender, erect; branchlets filiform; leaves loosely imbricate, short, broadly lanceolate-acuminate, subfalcate or more or less incurved, costate; strobiles ovate-globose, small.

The fragments referred to this species are merely two short branchlets, pl. iii, fig. 8, *loc. cit.*, and some indistinct cones imbedded into the stone. The leaves appear to be of the same form as those of this species figured by Heer, "Molet. Fl.," pl. i, fig. 10, generally a little broader and shorter, and the cones have the same character as that of fig. 12 of the same plate. These fragments are also comparable to the species as figured in Heer, "Fl. Arct.," iii, pl. xxvii, figs. 5 and 6, of the Upper Cretaceous of Atane. Professor Heer says, in the first description of this species, "Molet. Fl.," *l. c.*, that the leaves do not seem to have any medial nerves, and in "Fl. Arct.," *loc. cit.*, he remarks on the difference of the species from *S. rigida* by the absence of a medial nerve. As the costa is distinct on the leaves of the Dakota Group the relation is doubtful.

Hab.—Kansas, Clay Centre. *H. C. Towner.*

Sequoia condita, Lesqx.

Plate I, Figs. 5, 7, 9.

Hayden's "Ann. Rep.," 1874, p. 335, pl. iv, figs. 2–7.

Branches rigid, pinnately divided; branchlets slender, filiform, oblique; leaves short, oblong, thick, not pointed, narrowed to the decurring base, appressed to the stem, sometimes longer linear-acuminate, curved inward, nerveless; male ament oval, scaly, rhomboidal, apiculate.

This species is not rare in the shale of the Dakota Group, but as yet it has been found always imbedded into the shale and in small fragments, so that its characters cannot be stated with precision. Generally the leaves are lineal-oblong, acute, appressed to the stem, variable in length, sometimes longer, curved inward, resembling those of *S. fastigiata*, the medial nerve being indistinct. The cone, fig. 9, found upon specimens with branches of the species, is apparently an unopened fruiting catkin of this species. It has a slender short pedicel covered with very small obtuse scale-shaped leaves.

Hab.—Kansas; not rare.

GLYPTOSTROBUS, Endl.

Glyptostrobus gracillimus, Lesqx.

Plate I, Figs. 6–6b.

"U. S. Geol. Rep.," vi, p. 52, pl. ii, figs. 8, 11–11f; Hayden's "Ann. Rep.," 1874, p. 337.

I have figured here a mere fragment which I consider referable to the species, though the branch is a little thicker and the leaves ovate, somewhat like those of *Sequoia condita*, but shorter, as may be seen in comparing both figs. 5a and 6a.

The leaves of this plant and their disposition are remarkably similar to those of *Cyparissidium gracile*, Heer, "Fl. Arct.," iii, p. 74, pl. 19, fig. i, found at Kome and Atane.

THUITES, Sternb.

Thuites crassus, sp. nov.

Pinnately branching; branches comparatively thick, alternate; branchlets short, obtuse; leaves thick, broadly oblong, equilateral, as broad as long, closely imbricate in four rows; medial nerve distinct, inflated on the back.

Species closely allied to *Thuites Meriani*, Heer, "Fl. Arct.," iii, p. 73, pl. xvi, figs. 17, 18, differing especially by the great thickness of its branches, the leaves larger, broader, the facial and lateral of the same size.

Hab.—Seven miles northeast of Glasco. *Chs. Sternberg.*

PINUS. Linn.

Pinus Quenstedti, Heer.

Plate I, Figs. 3, 4.

Hayden's "Ann. Rep.," 1874, p. 336, pl. iii. figs. 6, 7.

Leaves in fassicles of five, very long and slender, thread-like, deeply nerved, the base inclosed in long cylindrical sheaths; cone cylindrical, very long; scales with broad rhomboidal shields (apophyses), acute on the sides, mammillate in the center.

The specimens representing this species are numerous but all fragmentary. The leaves are generally scattered and imbedded close together, their point of attachment by five is marked by the long sheaths forming deep holes into the stone; but none has been thus far found preserved entire. The species may be, therefore, different from that of Heer, described as above, and figured in "Molet. Fl.," p. 13, pl. ii, figs. 5–9. The thread-like long leaves, the long cylindrical cone, and the shields of the scales are, however, so much alike that I have scarcely any doubt on the identity of the Dakota Group species with that of Europe. The length of the leaves as given by Heer, who has had splendid specimens for description, is 20 centimeters. The fragments I have seen are 5 to 8 centimeters. The cylindrical cone, 22 millimeters broad, gradually tapering to the base, appears to be very long, its impressions perforating large stones, being at least 15 centimeters long. These cones are generally curved as in fig. iii. Heer represents them straight but of the same length and width.

In the "Flora of Gelinden" by Saporta and Marion, the authors remark (p. 19) that this fossil species does not differ by any important character from the living Mexican Pines with quinate leaves which now compose the section of the *Pseudo-strobus*.

Hab.—Near Fort Harker and Clay Centre, Kansas. *Chs. Sternberg* and *H. C. Towner*.

FRAGMENTS OF CONIFERS OF UNCERTAIN RELATION.

Abietites Ernestinæ, Lesqx.

"U. S. Geol. Rep.," vi, pl. i, fig. 7.

Sequoia formosa, Lesqx.

"U. S. Geol. Rep.," vi, pl. i, figs. 9, 9a.

Inolepis? species.

Plate I, Figs. 8–8c.

Hayden's "Ann. Rep.," 1874, p. 337, pl. iv, fig. 8.

Nutlets small, globular, short-mucronate, sessile upon slender branches.

c F 3

The specimen, fig. 8, shows the impression of three unopened globular, naked nutlets, which, as seen in figs. 8a and 8c, appear to contain small seeds which, in fig. 8c, are obcordate and inflated. These three last figures are all spread upon the same specimen with fig. 8.

The relation of this fragment to *Inolepis* is not certain. The fruits found mixed with a mass of decayed and broken remains of conifers may be considered as indeterminable, even in their generic relation, until better specimens are obtained.

Hab.—Dakota Group of Kansas.

MONOCOTYLEDONES.

GLUMACEÆ.

PHRAGMITES, Adans.

Phragmites cretaceus, Lesqx.

"U. S. Geol. Rep.," vi, p. 55, pl. i, figs. 13 and 14; pl. xxix, figs. 7, 7a.

Leaves and culms in fragments of various sizes; leaves lanceolate, blunt at the apex, doubly nerved; primary nerves thick or inflated under the thick epidermis, under which the intermediate veinlets, three or four, are discernible.

The fragmentary state of the first specimens found afforded reasonable doubt of their reference to this genus. But remains of plants of analogous character have been since discovered in the Upper Cretaceous of Greenland and described as *Arundo Grönlandica*, Heer, "Fl. Arct.." iii. p. 104, pl. xxviii, figs. 8–11. In this species the primary nerves do not appear separated by any intermediate veinlets, at least none could be observed by Professor Heer. This difference, and also the great size of the leaf, pl. xxix, fig. 7 of the Rep., *l. c.*, evidently separate the species.

DIOSCOREÆ.

DIOSCOREA, Plum.

Dioscorea? cretacea, Lesqx.

"U. S. Geol. Rep.," vi, p. 66, pl. xxviii, fig. 10.

PALMÆ.

FLABELLARIA, St.

Flabellaria? minima, Lesqx.

"U. S. Geol. Rep.," vi, p. 56, pl. xxx, fig. 12.

DICOTYLEDONES.

MYRICACEÆ.

MYRICA, Linn.

Myrica obtusa, Lesqx.

"U. S. Geol. Rep.," vi, p. 63, pl. xxix, fig. 10.

Myrica Dakotensis, Lesqx.

Plate IV, Fig. 9.

M. cretacea,[1] Lesqx., Hayden's "Ann. Rep.," 1874, p. 339, pl. iii, fig. 4.

Leaves narrowly lanceolate or lineal-oblong, gradually narrowed to a thick short petiole, crenulate on the borders; medial nerve flat and broad; lateral nerves at an acute angle of divergence, parallel, variable in distance, camptodrome; tertiary veins short, anastomosing with the secondary ones by nervilles at right angles.

The substance of the leaves is thick, coriaceous, the surface polished, the borders slightly reflexed and crenulate; the upper end of both the leaves representing the species is destroyed, but on fig. 9 the apex seems rounded or obtuse. They are 7 to 8 centimeters long and about 1½ centimeters broad in the middle.

The relation of these leaves is more distinctly marked with *M. Schenkiana*, Heer, "Quedl. Fl.," p. 11, pl. iii, fig. 1, and less distinctly with *M. cretacea*, Heer, *ibid.*, p. 10, pl. iii, figs. 2a, b, c. They are more lineal, the borders less deeply and more closely denticulate, the veins closer, etc. By their hard texture and their nervation the leaves are also comparable to those of some tropical species of *Salix*.

Hab.—Fort Harker, Kansas. *Chs. Sternberg.*

Myrica Sternbergii, sp. nov.

Leaves long, linear-lanceolate; borders distantly obtusely dentate; lateral veins at acute angles of divergence, comparatively thick, flexuous, simple or forking above the middle, the divisions entering the teeth in curving along the borders.

The specimen represents only a fragment of a leaf whose upper and lower parts are destroyed. The fragment is 7 centimeters long and 3 broad; the secondary veins or their primary divisions enter the teeth by

---

[1] *Myrica cretacea*, Heer, was, perhaps, published in the "Flora of Quedlinburg" before I described my species under the same name. But that work of Heer was then unknown to me. I change name, not being certain who has priority for it.

their ends or by anastomosing branches, diverging under the teeth and following the borders. By the size of the leaves and the distant obtuse teeth this species is related to *M. Thulensis*, Heer, "Fl., Arct.," iii, p. 107, pl. xxxi, fig. i; also closely allied by the nervation to *M. apiculata*, Sap. "Sézanne Fl.," p. 342, pl. iv, fig. 5.

Hab.—Two and a half miles north of Glasco, Kansas. *Chs. Sternberg.*

Myrica? semina, Lesqx.

"U. S. Geol. Rep.," vi, p. 63, pl. xxvii, figs. 4, 4a.

BETULACEÆ.

BETULA, Tourn.

Betula Beatriciana, Lesqx.

"U. S. Geol. Rep.," vi, p. 81, pl. v, fig. 5; pl. xxx, fig. 4.

Betulites denticulatus, Heer.

"Phyll. Crét. du Neb.," p. 15, pl. iv, figs. 5, 6.

Leaves short, ovate, denticulate, rounded at base; lateral nerves diverging in acute angle, craspedodrome, straight.

The craspedodrome nervation relates these leaves to *Betula* or *Alnus*, though the form of the leaves recalls the type of *Populus*.

Phyllites betulæfolius, Lesqx.

"U. S. Geol. Rep.," vi, p. 112, pl. xxviii, figs. 4, 7.

ALNITES, Goepp.

Alnites grandifolius, Newby.

"Notes on Ext. Fl.," Catal., p. 9; "Illustr.," pl. iv, fig. 2.

The species is not described by the author. The figure represents a large, round-oval leaf, narrowed to the petiole; the borders deeply regularly undulate, the lateral nerves at an acute angle of divergence, much branched on the lower side, craspedodrome like all the divisions.

The leaf represents the type of *Alnus glutinosa* by its form and size; but the borders are entire, merely undulate, not denticulate. The base of the leaf is more acutely narrowed to the petiole than it is generally in the leaves of this genus.

Hab.—Nebraska. Dr. *F. V. Hayden.*

CUPULIFERÆ.

FAGUS, Tourn.

Fagus polyclada, Lesqx.

"U. S. Geol. Rep.," vi, p. 67, pl. v, fig. 6.

Fagus cretacea, Newby.

Plate II, Figs. 6, 6a.

Newby., "Notes on Ext. Fl.," p. 23; "Illustr.," pl. ii, fig. 3.

Leaf oval, entire, slightly narrowed to the petiole; lateral veins sharply defined, numerous, parallel, craspedodrome, the points of the nerves being prominent and the intervals between them forming shallow sinuses.

To offer a point of comparison between this leaf and the one described as *Fagus polyclada* I have figured it again, distinctly tracing the nervilles, which are not visible on the original figure. The type of venation is that of *Fagus* not of *Rhamnus;* the marginal veinlets only are more distinct than in *F. ferruginea,* following the border in short curves anastomosing to the upper nervilles, nearly as in *Castanea.*

Hab.—Smoky Hill, Kansas. Dr. *F. V. Hayden.*

DRYOPHYLLUM, Debey.

Leaves lanceolate or oblong, generally dentate, penninerve; secondary nerves (in denticulate leaves) sub-opposite, straight, simple, entering the teeth directly by their points, or more rarely branching quite near the borders, one of the divisions entering a tooth, the other curving under, following the margins in wavy flexures and joining the next vein above. In the entire leaves the secondary nerves are more or less curved, camptodrome, with nervilles transversely decurrent, simple or forking, united by venules at right angles. (Sap.)

Saporta in describing the genus compares the species referred to it to some kinds of Oaks and Chestnuts with coriaceous leaves, now inhabiting the mountains of Asia and of Mexico, and which seem to have been the ancestors of the Oaks and Chestnuts of the present Flora of North America.

Dryophyllum (Quercus) primordiale, Lesqx.

"U. S. Geol. Rep.," vi, p. 64, pl. v, fig. 7.

Dryophyllum (Quercus) latifolium, Lesqx.

Plate IV, Figs. 1, 2.

Hayden's "Ann. Rep.," 1874, p. 340, pl. vi, fig. 1.

Leaf large, broadly ovate, rounded at base, deeply sinuate, obtuse or blunt at the apex; medial nerve thick; secondary nerves distant, straight or slightly curving up to the borders, the lower more or less branching.

The fine leaf, fig. 1, is nearly 12 centimeters long and 9 broad in the middle, its widest part, coriaceous, deeply undulate. The nervation is thick and coarse, the secondary nerves, 8 pairs, alternate, diverging at an angle of 50° are parallel, except a pair of basilar, thin, short marginal veinlets which, nearly at right angles to the medial nerve, follow close to the borders. The lower veins are more or less branching and enter the undulations or broad obtuse teeth, somewhat less prominent at the ends of the branches than at those of the nerves. The nervilles thin, but very distinct, are flexuous, at right angles to the veins, more generally continuous. Fig. 2 is an incomplete fragment which I consider as representing the same species. The upper end of the veins and of their branches are connected by strong nervilles following close to the borders; but they are not subdivisions of the secondary veins. This nervation is like that of *Castanea* and of some species of *Quercus*.

Hab.—Fort Harker, Kansas. *Chs. Sternberg.*

Dryophyllum (Quercus) Holmesii, Lesqx.

Plate IV, Fig. 8.

Dryophyllum (Quercus) salicifolium,[1] Lesqx., Hayden's "Ann. Rep.," 1874, p. 340, pl. viii, fig. 2.

Leaf linear-lanceolate, rounded in narrowing to the base, minutely acutely denticulate; lateral nerves numerous, parallel, alternate or opposite, slightly bowed subcamptodrome.

The fragment represents a slightly falcate somewhat thick leaf, rather membranaceous than coriaceous, with a narrow medial nerve and close parallel secondary ones, some of them as far as can be seen ascending to the teeth and passing under the sinuses by an upper branch, some others curving along the borders and reaching the teeth by short branchlets. This species is related to *Dryophyllum lineare*, Sap., "Séz. Fl.," p. 350, pl. iv, fig. 6. The teeth, however, of the American species are more distinct, turned outside in the lower part of the leaf, inclined upward in the upper part, as in *D. subcretaceum* of the same author, *ibid.*, p. 348, fig. 10.

Hab.—Near the San Juan River, at a higher Cretaceous stage than that of the Dakota Group; Southwest Colorado. *W. H. Holmes.*

---

[1] Name preoccupied as *Quercus salicifolia*, Newby., "Ext. Fl.," p. 24.

QUERCUS, Linn.

Quercus Dakotensis, sp. nov.

Leaf subcoriaceous, ovate-lanceolate, narrowed in rounding to the base, less abruptly, however, to an acute or blunt apex (not distinct), entire on the borders toward the base, nearly regularly dentate from below the middle upward, short pedicellate; medial nerve straight; secondary nerves thin, slightly bowed, divided into two or three branches, each entering a tooth.

The leaf is 9 centimeters long and 4½ centimeters broad in the middle; the point not distinct appears blunt; the pedicel is slender, nearly 1 centimeter long as far as it is seen before entering the stone; the secondary nerves diverge 55° to 60°.

The species is related to *Quercus Beyrichii*, Ett., "Kreidefl. von Nieders.," p. 14, pl. ii, fig. 2, from which it differs by the teeth not being turned upward or serrate, but abruptly acuminate outward; by the texture, which is not distinctly coriaceous; by thin secondary nerves and a narrow straight midrib. The upper veins are under the same angle of divergence, craspedodrome; the lowest pair, attached a little above the base of the leaf, follows the entire border up to the lower teeth. This species has also a degree of affinity to *Castanea Hausmanni*, Dkr., "Paleont.," iv, p. 181, pl. xxxiv, fig. 1. The teeth are of the same character.

These three species may be referable to the preceding genus, but the tertiary divisions of the veins are not discernible in any of them.

Hab.—South of Fort Harker. *Chs. Sternberg.* No. 62, Mus. Comp Zool., Cambridge.

Quercus hexagona, Lesqx.

"U. S. Geol. Rep.," vi, p. 64, pl. v, fig. 8.

This leaf, to which I could not indicate any related form when I described it (*l. c.*), is, in shape especially, allied to the Oligocene *Quercus Osbornii*, pl. xxxviii, fig. 17, which, itself, is comparable to *Quercus tephrodes*, Ung., as figured in Sieber, "Nord-Böhm Braunkohl. Fl.," iii, fig. 17.

Quercus Ellsworthiana, Lesqx.

"U. S. Geol. Rep.," vi, p. 65, pl. vi, fig. 7.

Another specimen referable to this species, as yet insufficiently represented and described, is a leaf of the same size and form as that of the "U. S. Geol. Rep.," *l. c.* The nervation is of the same character, at least

for the distance and the ramification of the secondary nerves; the lower
ones only are more open and more bowed in passing to the borders, the
lowest pair being nearly at right angles to the thick medial nerve. The
specimen is No. 1175 of the U. S. National Museum.

Quercus pornnoides, Lesqx.

"U. S. Geol. Rep.," vi, p. 66, pl. xxx, fig. 9.

The generic relation of this fragment, like that of the preceding, is
not positively ascertained.

Quercus Morrisoniana, sp. nov.

Plate XVII, Figs. 1, 2.

Leaves of medium size, coriaceous, petiolate, ovate-lanceolate, acuminate; medial
nerve strong; secondary nerves numerous, alternate, curved in passing to the borders,
camptodrome, simple, or some of them forking near the entire borders.

The species is related by its characters, shape, size, facies of the leaves,
and nervation to the Miocene *Quercus neriifolia,* A. Br. The midrib is
strong, prolonged into a petiole 1½ centimeters long. The lower veins are
slightly more open than the upper; all are nearly parallel, variable in
distance, more or less bowed in passing to the borders, which are very
entire. The leaves average 10 to 12 centimeters long, 3 to 3½ centimeters
broad in the middle where they are the widest, gradually narrowing in a
curve to the base and slightly decurring to the petiole.

The embedding material is a sandstone too coarse for the preservation
of the areolation; flexuous nervilles, transversely decurrent, are more or
less distinct. By this character the leaves are related to *Q. nervosa,* Sap.,
"Ét.," ii, i, p. 86, pl. iii, fig. 12.

Hab.—Base of the mountains, near Morrison, Colorado. *H. C. Beckwith.*

Quercus salicifolia, Newby.

"Notes on Ext. Fl.," p. 24; "Illustr.," pl. ii, fig. 1.

Leaves petiolate, smooth, thick, entire, abruptly pointed at both ends; medial
nerves strong, straight or flexuous; secondary veins unequal in size, strong near their
base, becoming finer, flexuous, and branching toward the borders, where some of them
inosculate by irregular curves while others terminate in the margin.

The facies of the leaf and the alternation remarked by the author of
large with smaller secondary veins, a character essentially pertaining to the
willows, seem to justify the reference of this leaf to *Salix.* The coriaceous

texture of the leaf and its smooth surface do not contradict this reference; for all the species of willows of the Dakota Group are coriaceous, as are generally the willows of the tropical or warm regions.

Hab.—Blackbird Hill, Nebraska. Dr. *Hayden.*

Quercus cuneata, Newby.

"Notes on Ext. Fl.," p. 25.

Leaves short, petiolate, lanceolate, pointed at both ends, acute, entire, or slightly wave-margined; midrib strong; secondary veins remote, nearly straight, with short intermediate ones; surface smooth, texture originally thick and coriaceous. (Ny.)

The author compares this species to *Q. imbricaria,* Michx., for the form and consistence of the leaves.

Hab.—Blackbird Hills, Nebraska. Dr. *Hayden.*

Quercus antiqua, Newby.

"Notes on Ext. Fl.," p. 26.

Leaves of medium size, lanceolate in outline, acute, often somewhat flexuous; margins serrate-dentate, with strong obtuse teeth, which are appressed or turned upward; midrib strong, percurrent; secondary veins numerous, of unequal strength, arched upward, craspedodrome. (Ny.)

Hab.—Lower Cretaceous sandstone, Banks of the Rio Dolores, Utah.

Quercus sinuata, Newby.

"Notes on Ext. Fl.," p. 27.

Leaves small, ovate in general outline, narrowed to the petiole or slightly decurrent; margins deeply lobed; lobes rounded, broader than the sinuses that separate them, three, nearly equal on either side; summit broadly rounded or obscurely lobed, often oblique; midrib straight or slightly flexed; secondary veins strong and simple, running to the margin of each lateral lobe. (Ny.)

The author compares the species to the living *Q. obtusiloba,* Michx.

Hab.—Same as the preceding.

SALICINEÆ.

SALIX, Tourn.

Salix nervillosa, Heer.

"Phyll. Crét. du Neb.," p. 15, pl. i, fig. 3.

Leaves oblong, lanceolate, very entire; secondary veins in an acute angle of divergence, curved, camptodrome; nervilles curved, at right angles to the midrib.

Hab.—Nebraska. Dr. *Capellini.*

Salix proteaefolia, Lesqx.
Plate I, Figs. 14–16; XVI, Fig. 3.

"U. S. Geol. Rep.," vi, p. 60, pl. v. figs. 1, 4.

The leaf figured (pl. xvi) is related to this species merely by its form, resembling that of pl. v, fig. 4, of the "Report." *l. c.* The nervation is indistinctly preserved, as in fig. 2 of the same plate. The other leaves (pl. 1, figs. 14–16) are all much narrower but broader toward the base, and gradually tapering to a long point. They have the same kind of venation and merely represent modified forms of this extremely variable species.

Hab.—Kansas, near Fort Harker.

Salix Meekii, Newby.
"Later Ext. Fl.," p. 19; "Illustr.," pl. i, fig. 1.

Leaves petioled, thin and delicate, lanceolate, acute at both ends, entire; midrib slender; secondary nerves fine, in an acute angle of divergence 35°, gently arched and anastomosing near the margins. (Ny.)

This is apparently the same species as the preceding, which, with an apparent difference in the texture of the leaves, the more or less acutely narrowed base, the great variety of size of the leaves, includes also the two following forms:

Hab.—Blackbird Hills, Nebraska. Dr. *Hayden.*

Salix cuneata, Newby.
"Later Ext. Fl.," p. 21; "Illustr.," pl. i, figs. 2, 3.

Leaves of medium size, sessile or short petiolate, lanceolate, acute at both ends, broadest toward the apex, gradually narrowed below to the base; medial nerve distinct; secondary veins delicate, with an acute angle of divergence (20°), gently arched above and inosculating near the margin. (Ny.)

The figures show the leaves larger in the middle, not toward the apex; they are more rapidly narrowed to the base and abruptly curve to the petiole in reaching it.

Hab.—Mouth of Sioux River, Nebraska. Dr. *Hayden.*

Salix flexuosa, Nowby.
"Later Ext. Fl.," p. 21; "Illustr.," pl. i, fig. 4.

Leaves narrow, linear, pointed at each end, sessile or very short petioled; medial nerve strong, generally somewhat flexuous; secondary veins diverging about 40°, somewhat branched and flexuous, curving and inosculating near the margins. (Ny.)

The author considers this as a variety of *S. Meekii.*

Hab.—Blackbird Hills, Nebraska. Dr. *Hayden.*

POPULUS, Linn.

Populus litigiosa, Heer.

"Phyll. Crét. du Neb.," p. 13, pl. i, fig. 2; Newby., "Illustr.," pl. iv, fig. 1.

Leaves round in outline, very entire at base; the two pairs of lower lateral veins opposite, the other alternate and distant; nervilles curved, simple or forking. (Hr.)

Hab.—Tekamah, Nebraska. Dr. *Capellini.*

Populus elliptica, Newby.

"Later Ext. Fl.," p. 16; "Illustr.," pl. iii, figs. 1, 2.

Leaves long-petioled, suborbicular or transversely elliptical, slightly cuneate at the base and apiculate at the summit; lower half of leaf entire, upper half or more very regularly and rather finely obtusely serrate or crenate, the points of the teeth inclining upward; primary nerves usually fine, sometimes three, radiating from the base at equal angles; from them the secondary veins spring at acute angles. (Ny.)

The species is remarkably similar, by the characters of the leaves, to *P. cuneata*, Newby., *loc. cit.*, p. 64, pl. xiv, figs. 1, 4, a Miocene species of the type of *P. arctica*, or is, perhaps, one of its numerous varieties.

Hab.—Blackbird Hills, Nebraska. Dr. *F. V. Hayden.*

Populus microphylla, Newby.

"Later Ext. Fl.," p. 17; "Illustr.," pl. iii, fig. 5.

Leaves very small, scarcely an inch in length, broadly cuneate at the entire base, rounded and deeply dentate from the middle upward; teeth conical, acute or blunt at the apex; nerves finely radiating from the base, branching above, the branches entering the teeth. (Ny.)

Hab.—Same as the preceding. Dr. *F. V. Hayden.*

Populus? cordifolia, Newby.

"Later Ext. Fl.," p. 18; "Illustr.," pl. iii, fig. 7.

Leaves heart-shaped, slightly decurrent on the petiole; margins entire; nervation fine but distinctly defined; medial nerve straight or slightly curved, running to the margin; lateral nerves, six on each side, diverging about 50°, nearly parallel, straight or slightly curved near the apex, the lower branching; nervilles at right angles or forking, rarely continuous. (Ny.)

Hab.—Same locality as the preceding. Dr. *F. V. Hayden.*

POPULITES, Lx.

Populites Lancastriensis, Lesqx.

"U. S. Geol. Rep.," vi, p. 58, pl. iii, fig. 1.

Populites elegans, Lesqx.

" U. S. Geol. Rep.," vi, p. 59, pl. iii, fig. 3.

Populites cyclophylla? Heer.

"U. S. Geol. Rep.," vi, p. 59, pl. iv, fig. 5.

From a remark of Professor Heer, this leaf is not referable to his *Populus cyclophylla*, as I supposed it. Indeed, from the craspedodrome nervation, this leaf is rather a *Cissites* than a *Populus*. Its relation is as yet undefined.

PLATANEÆ.

PLATANUS, Linn.

Platanus Newberryana, Heer.

"U. S. Geol. Rep.," vi, p. 72, pl. viii, figs. 2, 3; ix, fig. 3.

Platanus obtusiloba, Lesqx.

"U. S. Geol. Rep.," vi, p. 69, pl. vii, figs. 3, 4.

Platanus primæva, Lesqx.

"U. S. Geol. Rep.," vi, p. 69, pl. vii, fig. 2; xxvi, fig. 2.

Platanus Heerii, Lesqx.

Plate III, Fig. 1; VII, Fig. 5.

"U. S. Geol. Rep.," vi, p. 70, pl. viii, fig. 4: ix, figs. 1, 2.

Fig. 1 of pl. iii represents a fragment of a merely undulate, not lobate, leaf. The nervation has the normal character; the petiole is longer than I have seen it in any other specimen of this species. Another leaf, preserved entire, seen in the Museum of Comp. Zool., Cambridge, No. 225, is still smaller than this one, only 6 centimeters long and 5½ broad. It has the same nervation, the borders more deeply undulate, and two short, rather acute, lateral lobes. Fig. 5 of pl. vii is still a smaller form of this same species.

The leaves of this species have been found at two different localities on the Salina River and near Fort Harker, Kansas.

Platanus diminutiva, Lesqx.

"U. S. Geol. Rep.," vi, p. 73, pl. viii, fig. 5.

LIQUIDAMBAR, Linn.

Liquidambar integrifolium, Lesqx.

Plate XIV, Fig. 3.

"U. S. Geol. Rep.," vi, p. 56, pl. ii, figs. 1, 3; xxiv, fig. 2.

There is a degree of uncertainty in regard to the relation of the leaves described under this name, as I have remarked it in the "U. S. Geol. Rep.," *l. c.* If on one side they are related by their forms, especially the entire margin, to species of *Aralia*, or perhaps more of *Sterculia*, their nervation has more analogy to that of *Liquidambar* than to any other of the groups to which they have been compared. Two well-preserved specimens of the Museum of Comp. Zool., Cambridge, show the secondary veins somewhat variable in distance and divergence, moderately curving to quite near the borders, where they abruptly bend, following upward to the point where they anastomose in simple festoons. They are separated by short tertiary veins parallel to the secondary ones, dividing in the middle of the areas in joining the borders at right angles as nervilles. I have not observed this character in any of the fossil leaves which I have described as *Aralia*, nor do I find it in the few living species which I have for comparison. Another point of relation is remarked in the sub-cordate base of the leaves of the cretaceous species which, like *Liquidambar Styraciflua* and the common Miocene species *L. Europæum*, have the lower lateral lobes either curved back or at right angles to the petiole, so that the base of the leaf is never cuneate.

MOREÆ.

FICUS, Linn.

Ficus primordialis, Heer.

"Phyll. Crét. du Neb.," p. 16, pl. iii, fig. 1.

Leaves coriaceous, lanceolate, narrowed to the base, very entire, penninerve; lower pair of secondary veins at a very acute angle of divergence from the midrib, the others more open, all camptodrome.

I refer to this species two specimens (Nos. 26 and 33, Museum Comp. Zool., Cambridge), representing: the one, the impression of the upper surface of a lanceolate or oblong-lanceolate leaf, same size and shape as that of

Heer, with base and top also destroyed. The midrib is narrow, the secondary veins thin, the lower pair at a more acute angle of divergence; but the divergence of those above is gradually more obtuse, not abruptly so, as is the leaf of the "Phyllites." The veins are close, 5 to 6 millimeters distant, not decurring to the medial nerve, slightly arched in passing up toward the borders, which they follow in curves, anastomosing by nervilles at right angles with the anterior veins.

The other specimen bears the impression of the under surface of a leaf and the upper part of two others, these tapering into a long acumen. The areolation is very distinct, exposing a coarse reticulation composed of large, irregularly quadrate areas divided into small polygonal meshes.

These leaves have great affinity to those of the following species; they differ by their shape, oblong in the middle, by the secondary veins being more distinct, especially near and along the borders; the areolation appears to be of the same character.

Hab.—South of Fort Harker. *Chs. Sternberg.*

Ficus Halliana, Lesqx.

"U. S. Geol. Rep.," vi, p. 68, pl. xxviii, figs. 3, 9.

Ficus Beckwithii, sp. nov.

Plate XVI, Fig. 5; XVII, Figs. 3, 4.

Leaves sub-coriaceous, lanceolate or oblong, very entire, narrowed upward to a long acumen, more rapidly downward from below the middle and slightly decurring to the petiole; midrib strong, gradually thicker toward the base; secondary veins numerous, parallel, camptodrome; nervilles close, flexuous, and sub-continuous, at right angles to the veins.

This species is of the same type as *F. protogœa*, Heer, "Fl. Arct.," iii, p. 108, pl. xxx, figs. 1–8, differing by the form of the leaves, which in *F. protogœa* are oblanceolate or largest toward the apex. The veins are closer, though at the same angle of divergence, simple, 7 to 8 millimeters distant, at an angle of divergence of 35°. The nervilles also are very close and distinct. The leaves average 15 centimeters in length and 3½ to 4 centimeters broad below the middle.

Hab.—Near Morrison, Colorado. *H. C. Beckwith.*

Ficus? angustata, sp. nov.

Leaves narrowly lanceolate, comparatively long; medial nerve thick; secondary veins very close, at an acute angle of divergence, camptodrome.

The leaves are long, lanceolate, gradually acuminate, and also gradually narrowed to the base, 12 centimeters long, 2 centimeters broad. The veins at an angle of divergence of 20° pass upward slightly curved, and abruptly bend close to the borders following them in single bows. They are only 2½ millimeters distant.

This species differs from the preceding by narrower leaves, the more acute angle of divergence of the veins and their relative positions. It might be compared to *Rhamnus tenax*, Lesqx., "U. S. Geol. Rep.," vi, p. 109, pl. xxi, fig. 4; but the leaves, though of the same width, are nearly twice as long, and the angle of divergence of the veins is only half as broad; the medial nerve also is much thicker.

Hab.—Bluff Creek, Kansas. *Chs. Sternberg.*

Ficus magnoliæfolia, sp. nov.

Plate XVII, Figs. 5, 6.

Leaves very entire, oval or broadly lanceolate, broader below the middle, rounded in narrowing to the short petiole, and declined downward at the slightly decurring base; medial nerve of medium size, strict; secondary veins at an acute angle of divergence, close, very oblique, nearly straight from the midrib to near the borders, simply camptodrome.

The leaves, 8 to 10 centimeters long, 3½ to 5½ centimeters broad, with a short not inflated petiole about 1 centimeter long, appear somewhat thick but not coriaceous; they are acute or tapering to a short acumen (all the points are broken). The veins close, 5 to 7 millimeters distant, under an angle of divergence of 40°, are thin, parallel, except the lowest pair which is a little more oblique. The areolation is obsolete; only a few nervilles are seen at the end of the veins, anastomosing in marginal curves along the borders and close to them.

These leaves have a great likeness to those of *Magnolia Capellini*, Heer, "Phyll. Crét. du Neb.," p. 21, pl. iii. figs. 5, 6, differing especially by the more acute angle of the more numerous and closer secondary nerves. By this character, and also by the slightly decurring base of the leaves, they

are related to *M. alternans*, Heer, *l. c.*, p. 20, pl. iii, figs. 2, 4.   They may represent one of these species; but on account of the simple curves of the
veins close to the borders, and also of their position close to each other.
they do not appear to be referable to Magnolia.   The petiole is not inflated
as it is often in *Ficus*, but the lower pair of veins is more oblique, and, as
seen in fig. 6, the medial nerve is narrowly split or channeled in the middle.

Hab.—With the preceding. *H. C. Beckwith.*

Ficus Glascœna, sp. nov.

Leaves large, thick, coriaceous, polished on the surface, oblong-lanceolate, obtusely
pointed, narrowing and slightly decurring to the petiole; medial nerve very broad;
secondary veins thin, at a broad angle of divergence, scarcely curved in passing to the
borders, joining without curving to it a somewhat thick marginal vein.

The leaves are thick, 15 to 20 centimeters long, 6 to 7 broad; the
midrib 2 to 3 millimeters broad at base.   The type of venation resembles
that of *Ficus parasitica*, Shott., as figured by self-impression in "Bil. Fl.."
pl. xxiii, fig. 1; the thin lateral veins sometimes branching in the middle,
abruptly anastomosing to a somewhat thick marginal vein which follows
close to the borders in successive bows.   The secondary veins appear
separated by parallel thinner shorter tertiary veinlets; but the divisions
of the third order and the details of areolation are obscure.

Hab.—Two and a half miles south of Glasco, Kansas. *Chs. Sternberg.*

Ficus distorta, Lesqx.

Plate XIV, Fig. 4.

Hayden's "Ann. Rep.," 1874, p. 342, pl. v, fig. 6.

Leaf coriaceous, entire, obovate, unequilateral, pointed or acuminate, apparently
gradually narrowed to the base; nervation pinnate; secondary nerves thick, parallel,
equidistant, camptodrome; nervilles strong, at right angles to the veins, anastomozing
and subdividing into an irregularly quadrate or polygonal areolation.

A mere fragment of a leaf of which the upper and lower parts are
destroyed.   The characters do not positively indicate its relation to *Ficus*.
It is figured and described for future comparison.

Hab.—Near Fort Harker, Kansas.

Ficus laurophylla, Lesqx.

Plate I, Figs. 12, 13.

Hayden's "Ann. Rep.," 1874, p. 342, pl. v, fig. 7.
Laurophyllum reticulatum, Lesqx., "U. S. Geol. Rep.," vi, p. 76, pl. xv, figs. 4, 5.

Leaves coriaceous, polished on the upper face, entire, narrowly lanceolate, acuminate, gradually tapering to a short thick petiole; medial nerve thick, grooved on the upper side; secondary veins close, very open.

A large number of specimens of this fine species have been examined. Though generally more or less fragmentary and often erased on the surface, the essential characters may be generally recognized. The leaves vary in size from 10 to 20 centimeters long and from 1½ to 4½ centimeters broad in the middle. They are lanceolate, gradually narrowed both ways from the middle. The secondary nerves are parallel, unequal in distance, nearly at right angles to the midrib, and also nearly straight in passing to near the borders, where they curve and anastomose in festoons. They are generally separated by one or two tertiary veins attached to them by branches either oblique or at right angles, whose subdivisions compose an irregularly quadrate areolation.

By its nervation this species has a typical relation to *F. Glascana*. The curves of the secondary veins, which follow close to the borders in successive bows, form a kind of margin, as in the preceding species; but the veins distinctly curve to the festoons and compose them. They do not abruptly anastomose with them by their attenuated ends; for this reason the marginal flexures are thin, rarely distinct in this species, while in *F. Glascœna* they appear as formed by a truly independent nerve, more deeply and distinctly marked than the ends of the secondary veins.

In the collection of the Museum of Comp. Zool. of Cambridge I have found fourteen specimens of leaves same size and form as those described here, with the same character of areolation, but with the secondary veins at an acute angle of 30°. All the specimens are from the same locality, Elkhorn Creek, and seem to represent a truly different species. But the lateral veins and their divisions are not distinct enough to be satisfactorily described.

At first I considered the relation of these leaves to be with the *Laurineœ*. But as remarked already in the first description of this species,

c F 4

the venation is of the same type as that of some species of *Ficus* of both the present and the older floras, comparable, for example, to that of *Ficus Geinitzii*, Ett., "Fl., Niedersch.," p. 16, pl. ii, figs. 7, 9–11.

The two leaves figured, pl. i, show the under face, where the veins are more distinct and the medial nerve half-round. On the upper face the midrib is deeply channeled, but not inflated at the point of union to the short petiole which is rarely longer than 1½ centimeters.

Hab.—Commonly found throughout the Dakota Group formation from Minnesota to Southern Kansas.

PROTEACEÆ.

PROTEOIDES, Heer.

Proteoides daphnogenoides, Heer.

"U. S. Geol. Rep.," vi, p. 85, pl. xv, figs. 1, 2.

Proteoides grevilleæformis, Heer.

"U. S. Geol. Rep.," vi, p. 86, pl. xxviii, fig. 12.

Proteoides lancifolius, Heer.

"Quedlinb. Fl.," p. 12, pl. iii, figs. 5, 6.

Leaves narrowly lanceolate, narrowed in the upper part, very entire.

Two specimens, Nos. 63 and 76, of the Museum of Comp. Zool. of Cambridge, seem referable to this species. The first is a fragment of a linear-lanceolate leaf narrowed upward to an inclined apparently obtuse point, 8 to 9 centimeters long, 11 millimeters broad in the middle, the base destroyed. The medial nerve is narrow, and the thin lateral veins, two of which are seen near the base, come out at a very acute angle of divergence and are soon effaced upward.

The other leaf is larger, 16 millimeters broad in the middle, 8 centimeters long, lanceolate, gradually equally narrowed both ways, obtuse at the apex. Its medial nerve is flat, somewhat broader, 1 millimeter near the base, wherefrom two lateral nerves ascend at a very acute angle of divergence (about 10°), and no other veins are distinct up to above the middle of the leaf, where a few alternate ones come out at a broad angle of divergence, curving up as in fig. 6 of Heer, *loc. cit.* Except that this leaf is slightly broader the characters are identical.

Hab.—Near Fort Harker, Kansas. *Chs. Sternberg.*

EMBOTHRITES, Ung.

Embothrites (?) daphnoides, Lesqx.

"U. S. Geol. Rep.," vi, p. 87, pl. xxx. fig. 10.

From the comparison of a number of well-preserved specimens of *Andromeda Parlatorii*, Heer, recently received from Kansas, I am disposed to consider this fragment as referable to this last species.

LOMATIA, R. Brown.

Lomatia? Saportanea, Lesqx.

Plate III. Fig. 8 (enlarged).

Hayden's "Ann. Rep.," 1874, p. 346.
Tulba Saportanea, Lesqx., "U. S. Geol. Rep.," vi, p. 48, pl. xxix, figs. 1–4.

Leaves compound, linear in outline; ultimate divisions membranaceous or sub-coriaceous, narrowly lanceolate, acute, connate by the decurring base forming a narrow nerved wing to the rachis; medial nerve strong and straight, continuous to the apex; secondary veins simple, close, parallel, diverging at an acute angle in passing up close to the borders, which they follow in simple bows; tertiary veins shorter, anastomosing with the secondary ones by oblique diversely inclined veinlets.

The ultimate divisions of the leaves are parallel-oblique or somewhat curved downward, alternate or sub-opposite, a disposition similar to that of the divisions of the pinnæ of a number of species of ferns. They are gradually decurrent on the rachis, following it downward as a narrow-veined or smooth margin. The venation of the leaves is distinctly seen on the enlarged fragment, fig. 10.

My first impression in regard to these remarkable and fine vegetable remains was that they represented an extinct kind of fern. I even supposed that, considering the peculiar disposition of the leaflets and their venation, which is sometimes mixed with curved lines, we had here vegetable remains of a new type, constituting a link of transition between the ferns and the plants of a higher order. The segmentation of the leaves is similar to that of some species of fossil ferns, *Sphenopteris desmomera,*[1] for example, which, according to the remarks of the author, has no relation to any living fern; also related to the fragments described by Debey and Ettingshausen[2] under the generic name of *Monheimia.* For not only have they a similar division of the pinnæ, but, as seen in fig. 6, the nervation

[1] Saporta, "Plantes fossiles des lits à poissons de Cérin, p. 22, pl. xiv.
[2] "Urweltlicher Acrobryen," p. 31, pl. iv, figs. 1–10.

is somewhat analogous, the numerous parallel secondary veins curving up along the borders, some of them united by oblique veinlets.

Competent observers in Europe have contradicted these views and referred the fossil fragments to the *Proteacea*, comparing them to some species of *Lomatia;* and later I have received from the Oligocene of Florissant a large number of specimens, partly figured (pl. xliii), whose relation both with the Cretaceous species and with living specimens of *Proteacea* is evident.

Lomatia Saportanea, var. longifolia.
Leaves larger, divisions longer and broader.

None of the lateral leaflets are preserved entire, but from the fragments they are at least 8 or 9 centimeters long, though comparatively narrow, only ½ to 1 centimeter broad. The upper leaflets, two pairs of which are preserved, with the terminal upon one of the specimens, are 6 centimeters long and 7 millimeters broad, the terminal having the same size and characters.

Besides the difference in the size of the leaflets, these appear a little more distinctly coriaceous, and their surface is smooth without any trace of venation. Better specimens may prove this to be a different species.

Hab.—The specimens from which the variety is described are from Morrison, Colorado, procured by *A. Lakes.* The others, first described, are from Kansas.

LAURACEÆ.

LAURUS, Linn.

Laurus Nebrascensis, Lesqx.
"U. S. Geol. Rep.," vi, pl. 74, p. x, fig. 1; pl. xxviii, fig. 14.

Laurus macrocarpa, Lesqx.
"U. S. Geol. Rep.," vi, p. 74, pl. x, fig. 2.

Laurus proteæfolia, Lesqx.
Plate III, Figs. 9, 10; XVI, Fig. 6.
Hayden's "Ann. Rep.," 1874, p. 342, pl. v, figs. 1, 2.

Leaves subcoriaceous, broadly lanceolate, gradually narrowed from below the middle into a long blunt acumen, more rapidly attenuated to the base; medial nerve straight or slightly curved; lateral nerves slender, camptodrome, parallel, except the lower pair slightly more oblique.

The leaves vary in size from 9 to 16 centimeters long and 2½ to 3½ centimeters broad at or below the middle. The secondary veins, distinctly curved in passing from the midrib to the borders, are more or less distant, rarely separated by shorter tertiary veins cut at right angles by strong nervilles, which are simple or anastomosing in the middle, the upper ones ascending to the borders. The areolation is not seen, the surface appearing punctulate or closely dotted by small areoles.

In my first description of this species, *l. c.*, I compared it to *Proteoides daphnogenoides*, from the shape of the leaves only. This affinity is distant. By the form of the leaves this species rather resembles *Ficus Krausiana*, Heer, and *F. Beckwithii*, described above. Its venation is that of *Laurus Nebrascensis*, from which it differs by the narrower medial nerve, the secondary veins more slender and more curved in passing to the borders, the prolonged point of the leaves, etc.

Hab.—Near Fort Harker, Kansas. *Chs. Sternberg.* Recently found at Morrison, Colorado, by *A. Lakes.*

Laurus? modesta, sp. nov.

Plate XVI, Fig. 4.

Leaves small, linear-oblong, cuneate to the petiole; midrib thick; secondary veins irregular in distance, camptodrome, following close to the borders in prolonged curves.

There is only a fragmentary specimen of a small, apparently linear-lanceolate leaf (point broken), whose relation is not positively ascertained. The nervation is like that of *Laurus primigenia*, Ung., in Sap. "Él.," 2, 1, p. 89, pl. iii, fig. 8, the lateral veins at about the same distance and oblique in the same degree, curving high and close to the borders; but no trace of areolation is distinct. This fragment is also related to *Myrtophyllum pusillum*, Heer, "Quedl. Fl.," p. 14, pl. iii, fig. 10, represented by a still smaller fragment of leaf, round at base, with secondary veins curved and following high along the borders.

Hab.—Near Morrison, Colorado. *H. C. Beckwith.*

PERSEA, Gærtn.

Persea Leconteana, Lesqx.

"U. S. Geol. Rep.," vi, p. 75, pl. xxviii, fig. 1.

Persea Sternbergii, Lesqx.

"U. S. Geol. Rep.," vi, p. 76, pl. vii, fig. 1.

CINNAMOMUM, Burm.

Cinnamomum Scheuchzeri? Heer.

"U. S. Geol. Rep.," vi, p. 83, pl. xxx, figs. 2, 3.

Professor Heer considers the reference of these leaves to *C. Scheuchzeri* as uncertain; for though the form of the leaves is much the same, the middle nerve is too thick for that species, especially toward the point. Saporta is also of opinion that the presence of *C. Scheuchzeri* in the Cretaceous is very improbable, as in Europe this species is essentially in the upper Miocene. In his paper ("Descriptions of the fossil plants collected by Mr. George Gibbs"), Professor Newberry doubtfully refers to *Cinnamomum Heerii*, Lesqx., some leaves whose affinity of nervation is in his opinion with *C. Scheuchzeri* or *C. lanceolatum*. Following Professor Heer's opinion, I had changed the original name of the "Rep.," *l. c.*, to that less definite of *Daphnogene cretacea* (Hayden's "Ann. Rep.," 1874, p. 343); but if specific identification is not ascertainable from the fragmentary specimens obtained thus far, the close relation is at least indicated by the old name, which should, therefore, be preserved. Another reason against the change of name is the intimate relation, or perhaps identity, of the Cretaceous *C. Heerii*, with a Tertiary species of the genus.

Cinnamomum Heerii, Lesqx.

"U. S. Geol. Rep.," vi, p. 84, pl. xxviii, fig. 11.

Leaves thick, coriaceous, very entire, ovate, taper-pointed, rounded at the base to a short petiole; lateral nerves emerging a little above the base, ascending higher than the middle of the leaves, branching outside.

There is scarcely any modification to be made to the description of the "Rep.," *l. c.*, which I am able now to complete from a recently procured specimen of an entirely preserved leaf. This leaf, 9 centimeters long without the petiole (1 centimeter long), is broadest above the base, rounded to the petiole, joining it in an abruptly and short declining curve, and tapering above to a somewhat acute or merely blunt point. The medial nerve is broad and deep, enlarged to the base from the point of union of the lateral primary nerves 7 millimeters above the top of the petiole, gradually narrowed upward but distinct or persistent to the apex. The lateral nerves though thick are not as strong as the midrib, ascend in slightly curving inward up to nearly the second pair of secondary veins, where they are effaced near the borders. The secondary veins, two pairs, are alternate, distant, much curved in ascending high toward the borders, the lowest joining the medial nerve above the middle of the leaves, while from the base downward to the fork of the primary nerves the area is filled by a series of thin nervilles derived at right angles from the midrib. The lateral primary nerves are divided in numerous lateral branches, 5, 6 curving in passing outside toward the margins, where they become effaced.

This leaf is well enough represented by the figure in "U. S. Geol. Rep.," vi, made from a specimen whose borders had been ground from the middle downward and rounded to the point of union of the lateral nerves in such a way that the relative position of the nerves to the base of the leaf could not be ascertained, nor the disposition of the borders in joining the petiole. The size of the newly-found leaf is larger and its broadest point is close toward the base.

Excepting this last character, and its thinner venation, the Cretaceous leaves are very similar to those described from the Mississippi Eocene as *C. Mississippiense*, lately identified with numerous leaves of *C. affine*, of the Laramie and Carbon Groups. These are of about the same size, but all are rather oval-acuminate than ovate, the broadest part being in the middle. In *C. polymorphum*, to which both the Cretaceous and Tertiary species have been compared, the leaves are broader above the middle.

The specimen figured in "U. S. Geol. Rep.," vi, *l. c.*, came from near Ellsworth, Kansas. That of Nanaimo was, as far as I can recollect, in a still more imperfect state of preservation, and as I have not preserved a copy of the plates delivered to Dr. Evans, which have never been published. I am unable to see, if, indeed, the leaf of Nanaimo is identical with that of the Dakota Group. This, however, could not force a definite conclusion of the age of the flora of Nanaimo, as the Cretaceous type of *Cinnamomum* appears preserved with very little modification in the different Tertiary stages of this continent.

OREODAPHNE, Nees.

Oreodaphne cretacea, Lesqx.

"U. S. Geol. Rep.," vi, p. 84, pl. xxx, fig. 5.

A fine leaf of this species recently found in Kansas (No. 215, Coll. of the Museum Comp. Zool., Cambridge) has all the characters of the leaf figured. It differs merely by the secondary nerves not being as thick. The areolation is not distinct.

SASSAFRAS, Nees.

Sassafras Mudgei, Lesqx.
"U. S. Geol. Rep.," vi, p. 78, pl. xiv, figs. 3, 4; xxx, fig. 7.

Sassafras acutilobum, Lesqx.
Plate V, Figs. 1, 5.
" U. S. Geol. Rep.," vi, p. 79, pl. xiv, figs. 1, 2.

The form appears specific, as it is represented by leaves of very different size and always with the same characters. All the lobes are very entire, the lateral either broadly diverging, sometimes nearly at right angles to the midrib or erect; the venation is distinct but not coarse. The leaf, fig. 5, is one of the smallest seen of this species. The largest measures 12 to 14 centimeters long without the petiole, or still more, for I have seen from Kansas a fragment, only the middle lobe, 10 centimeters long from the sinuses to the apex and 4 centimeters broad. As the lateral lobes greatly vary in their divergence, of course the width of the leaves differ much. The species is especially abundant at Thomson Creek, near Fort Harker, with S. cretaceum and other forms of the same type.

SASSAFRAS (Araliopsis), Lesqx.

Sassafras (Araliopsis) cretaceum, Newby.
"Later Ext. Fl.," p. 14; " Illustr," pl. vi, figs. 1, 4 (fragments of leaves). Lesqx., "U. S. Geol. Rep.," vi, p. 80, pl. xi, figs. 1, 2; xii, fig. 2.

Sassafras (Araliopsis) obtusum, Lesqx.
S. cretaceum, var. obtusum, Lesqx., " U. S. Geol. Rep ," vi, p. 80, pl xii, fig. 3; xiii, fig. 1.

This form should be considered as specific, not merely on account of its shorter, more obtuse lobes, but particularly of the venation, which is much coarser than in the preceding species. The primary nerves, especially, are much broader and sharply cut. It is found with S. cretaceum at Thomson Creek; but it is also found by itself alone in other localities.

Sassafras (Araliopsis) mirabile, Lesqx.
" U. S. Geol. Rep.," vi, p. 80, pl. xii, fig. 1.
Platanus latiloba, Newby., " Later Ext. Fl.," p.24; " Illustr.," pl. ii, fig. 4.

To the characters indicated in "Rep.," vi, may be added the thick coriaceous substance of the leaves, which in small specimens appear horny; the great divergence of the lateral lobes nearly at right angles to the medial nerve and also generally curved down; the middle lobe is always comparatively short and broad.

Sassafras (Araliopsis) dissectum, sp. nov.

Leaves very large, long and narrowly cuneate to the petiole, palmately five-lobate by subdivision of the lateral lobes diverging at an acute angle from the medial one.

The leaves of this form are very large, some measuring 22 centimeters from the top of the petiole to the apex, 20 centimeters between the extremities of the lateral lobes. The base is narrowly cuneate, long, decurring to the petiole; the three primary divisions are joined in obtuse but narrow sinuses; the lateral ones at an acute angle of divergence are cut into two short obtuse dentate lobes, while the middle one is taper-pointed, not lobed, but deeply undulate-dentate. This form might be considered as a var. of *S. mirabile*, but it differs greatly in the general facies, the lateral lobes oblique erect lobed and unequilateral, the lateral primary nerves alternating at base or joined to the medial at a distance from each other, the long lanceolate undulate-dentate middle lobe and in the nervation, the primary nerves being thick indeed, while the secondary nerves and their branches are thin, generally effaced along the borders.

Hab.—This form has not been seen among the numerous specimens of fossil plants examined from the Dakota Group until recently. It is represented in the collection of the Museum of Comp. Zool. of Cambridge by a number of fine specimens, all obtained from 3 and 7 miles north of Fort Harker by *Chs. Sternberg.*

Sassafras (Araliopsis) recurvatum, Lesqx.

Platanus recurvata, Lesqx., " U. S. Geol. Rep.," vi, p. 71, pl. x, figs. 3-5.

Leaves three to five palmately lobed; lobes nearly equal in length, the medial broader; lateral nerves curving downward, either simple with mere secondary veins or forking above the base; lobes undulate or obtusely dentate on the borders.

This form is evidently transient in its characters. By the cuneate and decurrent base of the leaves joining the petiole at a distance below the point of union of the three primary veins and by the trilobate division, it is a *Sassafras.* But by the irregularity of the lobes or the subdivisions of the leaves in lobes and teeth, it seems referable to *Platanus,* while a tendency to become five-lobate by the forking of the lateral nerves is a character of the *Araliaceæ*. This last character is still more marked in the following species.

This form is very rare. Except the specimens figured in the "Rep.," *l. c.*, I have not seen any identifiable with it, except a well-preserved leaf. No. 148, counterpart 105, of the Museum Comp. Zool., Cambridge, which in all its characters, especially by its peculiar nervation, represents in a diminutive form fig. 3 of pl. x. The lateral nerves join the medial only a little above the base of the leaf, and the lower pair of secondary nerves follow upward along the borders and by an inward curve anastomose with the outside curved end of the second pair above the middle of the leaf.

Sassafras (Araliopsis) platanoides, sp. nov.

Plate VII, Fig. 1.

Leaves narrowly cuneate from the middle downward, palmately five-lobate in the upper enlarged part; lobes short, the upper half-round or obtuse, apiculate, the lower deltoid-acute; primary nerves tripartite from far above the base of the leaves; lateral nerves branching in the middle, primary and secondary divisions passing out to the points of the lobes.

The leaf figured is 13 centimeters long from the point where it joins the enlarged medial nerve in gradually decurring to it, and 11 centimeters broad between the lower lateral lobes, which, though shorter than the upper ones, are turned outside, while those above are directed upward; the point of union of the veins is 2½ centimeters above the base of the leaf, the medial nerve underneath being 3 millimeters thick or three times as broad as the medial nerve above the division. The lobes are of a peculiar shape, the lower ones deltoid-acute, short, about 1 millimeter long; the upper ones longer, rounded and narrowed to a blunt apex; the terminal is of the same shape but still longer; all are joined in obtuse sinuses.

The close relation of this leaf to *Platanus Heerii*, "U. S. Geol. Rep.," vi. pl. ix, figs. 1, 2, will be easily recognized; but still, the long narrowly wedge-form base, the subdivision of the lateral primary nerves, are characters represented in *Araliopsis*, especially in the preceding species, so that it is extremely difficult to say with which of these generic divisions this kind should be identified.

Hab.—Near Clay Center, Kansas. *H. C. Towner*, from a figure communicated. But other leaves of the same characters, only a little smaller (Nos. 694, 672, Museum Comp. Zool., Cambridge), have been found by *Chs. Sternberg*, on Thomson Creek, 7 miles south of Fort Harker.

Sassafras (Araliopsis) subintegrifolium, Lesqx.

"U. S. Geol. Rep.," vi, p. 82, pl. iii, fig. 5.

From a number of specimens more or less similar to those of the leaf figured "U. S. Geol. Rep.," vi, *l. c.*, I believe it represents only a deformation of *S. cretaceum*, especially of its variety *obtusum*. I have, however, received quite recently, from North Kansas, leaves of Sassafras perfectly entire or lobate on one side only, identical in shape and size with the leaves of *Sassafras officinale* commonly found also entire, bilobate or trilobate. They were sent by Mr. *L. C. Mason*, of Delphos.

ARISTOLOCHIACEÆ.

ARISTOLOCHIA, Tourn.

Aristolochia dentata, Heer.

"Phyll. Crét. du Neb.," p. 18, pl. li, figs. 1, 2; Lesqx., "U. S. Geol. Rep.," vi, p. 87, pl. xxx, fig. 6.

DIOSPYRINEÆ.

SAPOTACITES, Ett.

Sapotacites Haydenii, Newby.

"Later Ext. Fl.," Catal., p. 8; "Illustr.," pl. v, fig. 1.

No description is given of this species. The leaf, of medium size, is obovate, slightly emarginate at the obtuse apex; secondary nerves at an acute angle of divergence, close, curved in passing up toward the borders, divided by short oblique veins detached from both sides of the lateral nerves.

Hab.—Nebraska. Dr. *F. V. Hayden.*

DIOSPYROS, Linn.

Diospyros primæva, Heer.

"Phyll. Crét. du Neb.," p. 19, pl. i, figs. 6, 7; Newby., "Later Ext. Fl.," Catal., p. 8; "Illustr.," pl. iii, fig. 3.

Leaves oblong-oval, very entire, rather obtuse at the apex; secondary veins flexuous, branching, camptodrome.

The author compares it to his *D. anceps* of the European Miocene, and to *D. Alaskana* of the same formation of Alaska. The species is not rare in Kansas.

Diospyros ambigua,[1] Lesqx.

D. anceps, Lesqx., "U. S. Geol. Rep.," vi, p. 89, pl. vi, fig. 6.

Diospyros rotundifolia, Lesqx.

"U. S. Geol. Rep.," vi, p. 89. pl. xxx, fig. 1.

ERICACEÆ.

ANDROMEDA, Linn.

Andromeda Parlatorii, Heer.

"Phyll. Crét. du Neb.," p. 13, pl. i, fig. 5; Lesqx., "U. S. Geol. Rep ," vi, p. 83, pl. xxiii, figs. 6, 7; xxviii, fig. 15.

Andromeda affinis, Lesqx.

Plate 11, Fig. 5.

Hayden's "Ann. Rep.," 1874, p. 348, pl. iii, fig. 5.

Leaf thick, narrowly lanceolate, acuminate, entire; medial nerve comparatively thick; lateral veins close, parallel, at an acute angle of divergence, camptodrome.

The leaf, 5½ centimeters long, 11 millimeters broad in the middle, is gradually narrowed downward to the petiole and upward to a somewhat long acumen; the angle of the lateral nerves is 30°; the areolation is composed of round or quadrate polygonal minute areoles.

This species is closely allied to the preceding; the veins are less oblique and more curved.

Hab.—Spring Cañon, with fragmentary leaves of *A. Parlatorii.* Dr. *F. V. Hayden.*

ARALIACEÆ.

ARALIA, Linn.

Aralia formosa, Heer.

Plate XI, Figs. 3, 4.

Heer, "Moletein Fl.," p. 13, pl. viii. fig. 3.

Leaves petioled, triple-nerved, trilobate; lobes dentate, blunt at the apex.

This species, as represented by American specimens, though positively identified, presents a few unimportant points of difference. In Heer's figures the base of the leaves is wedge-form and the divisions oblique; in those which I have for examination the middle lobe is oval or lance-

---

[1] The name of this species is changed as preoccupied by Heer.

olate, the lateral linear lanceolate, not enlarged in the middle, as far
as seen from the one partly preserved, and the borders are obtusely serrate
from near the base. In Heer's figures the medial lobe is shorter and narrower, and it is, like the other, denticulate only in the upper part. The
secondary veins are not very distinct; a few, of which the base only is
seen, are parallel, close, at an open angle of divergence. The leaves are
thick; the petiole is not preserved, but as seen in Heer's specimen it is
short and thick.

Heer compares this species for the shape of the lobes to *A. Japonica*,
which, however, has the leaves five-lobed, and indicates its relation to *A.
primigenia* of Mount Bolca and of Alumbay.

Hab.—Near Morrison, Colorado. *H. C. Beckwith.*

Aralia Saportanea, Lesqx.

Plate VIII, Figs. 1, 2; IX, Figs. 1, 2.

Hayden's "Ann. Rep.," 1874, p. 350, pl. 1, fig. 2.

Leaves large, sub-coriaceous, triple-nerved and five-lobate by division of the
lateral nerves, fan-shaped in outline, narrowed in a curve or broadly cuneate, and
decurring to a long slender petiole; lobes narrowly lanceolate or linear-lanceolate,
acute or blunt at the apex, equally diverging, distantly dentate from below the middle
upward; secondary nerves sub-camptodrome.

This beautiful species is known by numerous finely preserved specimens. The leaves, 9 to 20 centimeters long from the top of the petiole to
the summit of the middle lobe, are of the same width between the points
of the lower lateral lobes; the petiole is long and comparatively slender,
though appearing thick upon one of the specimens, probably enlarged and
flattened by compression. The preserved broken part on one of the leaves
measures 5 centimeters. The lobes cut down to about two-thirds of the
leaves are narrowly lanceolate, slightly narrower near the obtuse sinuses,
equally diverging, the lower lateral ones much shorter, curved down, and
decurring to the base of the leaves. The leaves, triple-nerved from the
division of the primary nerves a little above the base, become five-nerved
from the forking of the lateral nerves at a short distance from their base.
The secondary veins emerge at an acute angle of 30°, curve in ascending
to the borders, and sometimes enter the teeth by their ends; the upper
more generally follows close to the borders in festoons, emitting under the

teeth short branches which enter them. There are not any intermediate
tertiary veins, but the nervilles are strong, often continuous, anastomosing
in the middle of the areas and forming by subdivisions a small quadrangular areolation (pl. viii, fig. 1). The typical relation of these Aralia
leaves is marked with *Sassafras (Araliopsis) cretaceum* and *S. mirabile*,
though the generic and specific characters are far different.

Hab.—South of Fort Harker. *Chs. Sternberg.* A number of splendid
specimens have been found all at the same locality near Brookville, Kansas.

Aralia quinquepartita, Lesqx.

"U. S. Geol. Rep.," vi, p. 90, pl. xv, fig. 6.

Of this species, described, *l. c.*, from two fragmentary specimens, I
have now seen some better leaves. One, the largest, is 16 centimeters
long from the top of the petiole to that of a lateral lobe preserved entire.
It is deeply divided into six narrow oblanceolate lobes, obscurely dentate
toward the apex, the lower lateral nearly entire. The medial lobe, 2 centimeters broad above the middle, is only 1 centimeter broad near the sinus.
Though somewhat thick, the leaves are rather membranaceous than coriaceous, the upper face smooth. The lateral veins are obsolete, appearing
very thinly distributed, about like those of *A. Saportanea*. The division
of this leaf in six is abnormal; the primary lateral nerves on one side
fork twice and therefore form three lobes, while on the other side the lateral nerves fork once only and have thus two divisions only.

Hab.—The best specimens seen of this form are from south of Fort
Harker. *Chs. Sternberg.*

Aralia Towneri, Lesqx.

Plate VI, Fig. 4.

Hayden's "Ann. Rep.," 1874, p. 349, pl. iv, fig. 1.

Leaf large, coriaceous, polished on the upper face, irregularly five-lobed to below
the middle; lobes entire, oblong, obtusely pointed; primary nerves in three, from near
the top of the petiole, the lateral ones forked at a distance from the base; secondary
veins open, variable in distance, very curved in passing toward the borders, camptodrome, separated by short tertiary veins parallel to them or at right angles to the midrib.

The leaves of this fine species are, as seen from another better preserved specimen, 15 centimeters long from the top of the petiole and 22 to
24 centimeters broad between the points of the lobes, which, descending

much lower than the middle, are 7 to 10 centimeters long and 3 to 3½ centimeters broad. The primary nerves are comparatively narrow; the form of the lobes is oblong, the point somewhat obtuse, the sinuses broad and also obtuse. The secondary nerves distant, nearly simple, at an open angle of divergence, pass toward the borders in curves and follow them in festoons, anastomosing by nervilles with those above. They are generally separated by short tertiary veins forming by ramifications in more or less oblique directions, square or polygonal, large meshes.

Hab.—Clay Centre, Kansas. *H. C. Towner.*

Aralia subemarginata, sp. nov.

Leaves of medium size, thick, coriaceous, five-palmate, cuneate to the base; lobes cut to the middle of the leaves, entire, obovate, rounded or emarginate at the apex; primary nerves in three, the lateral forking near the base; venation camptodrome.

The lobes of this leaf are nearly equal in length, about 5 centimeters long from the narrow obtuse sinuses, 5 to 6 centimeters broad in the upper part; lateral veins few, distant, 3 or 4 pairs, some of them forking on the lower side, much curved in passing to the borders. This species is closely allied to the preceding, differing by the short, obovate, rounded or emarginate lobes and the nervation. The only specimen seen is No. 810 of the Museum Comp. Zool., Cambridge.

Hab.—Three miles southeast of Fort Harker, Kansas. *Chs. Sternberg.*

Aralia tenuinervis, sp. nov.

Plate VII, Fig. 4.

Leaf small, truncate at base, palmately five-lobed; lobes much diverging, lanceolate or linear-lanceolate, acute; sinuses broad and obtuse; primary nerves thin, flexuous, apparently diverging from the same point near the base of the leaf; lateral veins close, parallel, camptodrome.

The base of the leaf is destroyed and the point of union of the lateral nerves is not seen. It appears to be about like that of fig. 3 of the same plate, a leaf related by its shape. The thin primary nerves, the close lateral thin veins, separate this species from all the others described above. Its type is that of *Aralia angustiloba*, Lesqx., of the Chalk Bluffs of the Gold-gravel formation of California.

Hab.—Clay County, Kansas. *H. C. Towner.*

Aralia radiata, sp. nov.

Plate VII, Figs. 2, 3.

Leaves small, palmately five-lobed; base truncate and abruptly declined to the petiole; lobes equally diverging, lanceolate-acuminate, the lower at right angles to the medial nerve; primary nerves in three or five united near the basilar border of the leaves.

This description and the figures of this species are made from sketches communicated by Mr. H. C. Towner, the discoverer. As I have seen a poorly preserved specimen only, apparently representing the species, I am unable to give more details on the characters. In fig. 2 the lateral nerves are branching a little above the base. This division is observed in most of the Cretaceous leaves I have described of this genus, and it is especially from this kind of nervation that I have considered them as referable to *Aralia*. But in fig. 3 the primary veins are in five from the base, and this is a character of *Sterculia*. The great similarity of the leaves cut to two-thirds of their length into lanceolate, gradually cuneate lobes, the habitat at the same locality, seem to prove that they represent the same species.

Hab.—Clay Centre, Kansas. *H. C. Towner.*

Aralia concreta, Lesqx.

Plate IX, Figs. 3, 4, 5.

Hayden's "Ann. Rep.," 1874, p. 349, pl. iv, figs. 2, 3, 4.

Leaves small, very thick, coriaceous, palmately five-lobed to below the middle, broadly cuneate and curving to the petiole; lobes linear or narrowly lanceolate, very entire; primary nerves three, from a little above the border base of the leaves, tho lateral forking, all thick, flat, and deep by impression, preserving nearly the same size to the top of the obtusely-pointed leaves.

The leaves vary in diameter from 5½ to 8 centimeters between the points of the lateral lobes, being shorter than broad. The secondary nervation and areolation are totally obsolete. Fig. 4 is a remarkable form. On account of the rounded base of the leaf the lobes are not as widely diverging and the sinuses narrower. The essential characters, great thickness of leaves, broad percurrent primary nerves, the size also being the same, the difference cannot be considered as specific.

Hab.—Clay Centre. *H. C. Towner.* Bluff Creek, Ellsworth County, Kansas. *Chs. Sternberg.*

HEDERA, Linn.

Hedera ovalis, Lesqx.

'U. S. Geol. Rep.," vi, p. 91, pl. xxv, fig. 3; pl. xxvi. fig. 4.

Hedera Schimperi, Lesqx.

Plate IV, Fig. 7.

Hayden's "Ann. Rep.," 1874, p. 351, pl. vii, fig. 5.

Leaf sub-reniform, broader than long, rounded at the top, abruptly narrowed or obliquely sub-truncate to the petiole, three-nerved from a little above the base; lateral nerves curving and more or less oblique toward the borders, anastomosing by thick branches and veinlets with the divisions of short distant secondary veins curving along the borders and entering by short veinlets the distant slightly marked denticulations of the margins.

The leaf is coriaceous, 6½ centimeters broad and 6 centimeters long without the petiole, which is only 7 millimeters long. As seen on the specimen it appears enlarged to a point of attachment, not very distinct, however. The lateral veins are inclined on one side toward the medial nerves; on the other they rather tend down or toward the borders; the veinlets all nearly at right angles, anastomosing with the divisions of the secondary veins, form an irregular areolation of angular, square, or polygonal meshes. The areolation is of the same character as in the preceding species, and is analogous to that of *Greviopsis tremulæfolia* and of *Cissus ampelopsidea*, Sap., and recognizable also in the following species.

Hab.—South of Fort Harker. *Chs. Sternberg.*

Hedera platanoidea, Lesqx.

Plate III, Figs. 5, 6.

Hayden's "Ann. Rep.," 1874, p. 351, pl. iii, fig. 3.

Leaf small, broadly ovate, rounded at the top, truncate at base, short petioled, entire, triple-nerved at a short distance above the basal borders of the leaves; primary nerves craspedodrome.

The leaves, five to six centimeters in diameter, are about as broad as long; the borders are entire, though somewhat forced outside over the points of the primary nerves and thus very obscurely and obtusely tri-lobate. The lowest branches of the primary lateral nerves follow the borders in festoons along the base of the leaves as in the preceding species, and there is also under the primary nerves a pair of marginal veinlets at

c f 5

right angles to the midrib. The secondary veins and their divisions all
reach to very near the borders, even seem to reach them, anastomosing at
their ends with a veinlet which follows close to the margins in successive
short curves like a marginal vein. The nervilles are strong, more or less
at right angles to the nerves, not continuous, anastomosing in the middle
of the areas, composing a net of large irregular quadrangular or polygonal
meshes. The surface of these leaves is rough, the venation deep and distinct, the substance thick, nearly coriaceous; the short petiole (7 millimeters long) is enlarged at the base.

Hab.—Near Fort Harker. *Chs. Sternberg.*

AMPELIDEÆ.

CISSITES, Heer.

Leaves more or less deeply trilobate by the extension of the lateral primary
nerves always in three, rounded and broadly cuneate to the base; lobes deltoid or
round, entire or dentate, sometimes lobed; secondary nerves mostly camptodrome.

Under the name of *Cissites insignis,* and without definition of the genus,
Professor Heer has described a fragment of leaf which has apparently a
degree of affinity to those which I place under this generic division. The
leaves are closely allied to *Araliopsis* by the primary nervation always
being trifid generally from a distance above the basal borders, and by the
areolation and the more or less distinctly trilobate division. The secondary veins are generally camptodrome.

Cissites insignis, Heer.

"Phyll. Crét. du Neb.," p. 19, pl. ii, figs. 3 (1 restored).

Leaves coriaceous, palmately deeply trilobate; lateral lobes very unequal, lobes
crenate at the apex.

This leaf is very coriaceous, triple-nerved, deeply palmately trilobate.
The lower part of the lower lobe is larger than the upper, which is entire
and bears three obtuse teeth toward the base; the secondary veins are
thin, anastomosing in curves at a distance from the borders.

Cissites salisburiæfolius, sp. nov.

Sassafras obtusum, Lesqx., "U. S. Geol. Rep.," vi, p. 81, pl. xiii, figs. 2, 4.
Populites salisburiæfolius, Lesqx., "Am. Jour. of Sci. and Arts," xlvi, 1868, p. 91.

These leaves, first described as *Populites,* then as *Sassafras* or *Arali-*

opsis, and now as *Cissites*, have indeed some characters which relate them to *Araliopsis*. They are palmately trilobate, have about the same form as *Araliopsis cretaceus* var. *obtusus*, and an analogous distribution of the nerves and secondary veins. They differ much by the thin texture of the leaves and the disposition of the lobes to become more or less obtusely and distinctly dentate at the apex, as seen by figs. 2 and 4. The rapidly narrowed base and the very long petiole give to them a peculiar fan-like shape. Their relation to this group seems indicated by their affinity to *Cissites insignis*.

Cissites Harkerianus, Lesqx.

Plate III, Figs. 3, 4.

Hayden's "Ann. Rep.," 1874, p. 352, pl. vii, figs. 1, 2.
Sassafras (Araliopsis) Harkerianum, Lesqx., "U. S. Geol. Rep.," vi, p. 81, pl. xi, fig. 4.

Leaves coriaceous, broadly rhomboidal in outline, and cuneate to the petiole, palmately sub-trilobed; lateral primary veins joined at a short distance above the base; secondary veins and their divisions camptodrome.

The leaves figured here are smaller than fig. 4, pl. xi, of the "U. S. Geol. Rep.," vi; but this is the only difference, and a number of specimens have been found of leaves of intermediate size. The nervation is, of course, more or less pronounced, according to the face exposed upon the stone. The relation of this and the preceding species to *Araliopsis* is easily remarked.

Cissites affinis, Lesqx.

Platanus affinis, Lesqx., "U. S. Geol. Rep.," vi, p. 71, pl. iv, fig. 4; xi, fig. 3.

Leaves coriaceous or sub-coriaceous, triple-nerved from near the base, sub-trilobate, rounded in narrowing to the petiole, broadly deltoid to the apex; borders marked by short distant teeth at the points of the excurrent nerves and their branches.

Cissites acuminatus, Lesqx.

Plate V, Figs. 3, 4.

Hayden's "Ann. Rep.," 1874, p. 353, pl. viii, fig. 1.

Leaves deltoid from the middle to the acute point, rounded from the middle downward to the petiole, triple-nerved from the base.

These leaves, 7 to 8 centimeters long and nearly as broad, much resemble those of the preceding species; they differ merely by the borders being entire, the secondary nerves more numerous and camptodrome. In fig. 4 the points of the lower pair of these lateral nerves reach to the borders

and force them outside, forming short teeth. The difference between this
and the preceding form becomes, therefore, less marked and may not be
considered of specific value. But the same remarks can be made on the
numerous transitional forms of this peculiar flora, as it has been remarked
already.

Hab.—Near Fort Harker. *Chs. Sternberg.*

Cissites, Heerii, Lesqx.

Plate V, Fig. 2.

Hayden's "Ann. Rep," 1574, p. 353, pl. v, fig. 3.

Leaf fan-shaped in outline, broadly cuneate to the base from above the middle,
divided at the upper border into five nearly equal acute lobes separated by broad
sinuses; primary nerves trifid from above the basal border of the leaf, ascending with
the lower pair of secondary nerves to the points of the teeth; upper lateral veins and
all the subdivisions camptodrome.

Though the base of the leaf is destroyed its outline is clearly defined
by the preserved part of the borders and the direction of the lateral primary veins. Except that the two lower secondary nerves ascend to the
points of two lobes, the nervation of the leaf is of the same type as that
of the two preceding species. Though the close relation of these leaves
is evident, this one cannot be compared to *Araliopsis*. It, therefore,
authorizes a separation of this group, which by its characters is related to
the *Ampelideœ*, especially to *Cissus*.

Hab.—Near Fort Harker, Kansas. *Chs. Sternberg.*

Ampelophyllum, Lesqx.

Hayden's "Ann. Rep.," 1874, p. 354.

Leaves ovate or obovate, obtuse, entire, narrowed to a long petiole, or sub-cordate,
palmately three-nerved from above the base; nerves flexuous, branching on both sides,
ascending to the borders.

Ampelophyllum attenuatum, Lesqx.

Plate III, Fig. 2.

Hayden's "Ann. Rep.," 1874, p. 354, pl. ii, fig. 3.

Leaf sub-coriaceous, cuneiform in outline, enlarged and rounded at the top;
borders entire, wavy; lateral primary nerves joining the middle at a distance from the
base, flexuous, branching out and inside, ascending to the borders.

The leaf, 6½ centimeters long without the petiole and about the same
width between the points of the primary lateral nerves, is rounded at the

FLORA OF THE DAKOTA GROUP.

top and undulate by the out-running of the veins. It is triple-nerved from a distance above the base, and has above the point of connection of the primary nerves two or three pairs of alternate secondary veins, variable in distance, straight or curved, unequally parallel, reaching the borders either directly or by their branches, which by oblique branchlets or by connections of nervilles at right angles form irregular quadrate large meshes. There are under the primary nerves two pairs of marginal veinlets with the same degree of divergence as the primary ones (40°–50°). The form of this fine leaf and its mode of nervation are peculiar, and of a character analogous to that of leaves described under the generic name of *Greviopsis* in the "Sézanne Flora" by Saporta. There is, however, a marked difference in the primary ternate nervation and in the entire borders of the leaves. The two lower pairs of tertiary veins show also for this leaf a relation to *Credneria*, and especially to the small leaf of *Platanus Heerii*, pl. iii, fig. i. The secondary and tertiary nerves are of a different character.

Hab.—South of Fort Harker. *Chs. Sternberg.*

Ampelophyllum ovatum, Lesqx.

Hayden's "Ann. Rep.," 1874, p. 355.
Celtis ? ovata, Lesqx., "U. S. Geol. Rep.," p. 66, pl. iv, figs. 2, 3.

Leaves ovate, obtuse or undulate, truncate or obtusely pointed, enlarged toward the base and abruptly rounded and sub-truncate or cordiform at base; nervation trifid from the base, craspedodrome.

Though the relation of these leaves to the preceding species is not very distinct, it is, however, more marked than to the leaves of *Celtis*. But for the craspedodrome, and especially the ternate primary nervation, they might be referable to *Populus* or *Populites*, having indeed some degree of affinity to *P. elegans*, Lesqx., "U. S. Geol. Rep.," vi. pl. iii. fig. 3.

HAMAMELIDEÆ.

HAMAMELITES, Sap.

Leaves membranaceous, glabrous, petiolate, oblong-lanceolate or ovate; nervation pinnate; secondary nerves at an acute angle of divergence, craspedodrome, branching on the lower side; branches and subdivisions generally camptodrome.

The leaves described in this generic division have the essential characters of the leaves of both *Hamamelis* and *Alnus*.

Hamamelites tenuinervis, sp. nov.

Leaf broadly ovate, rounded at both ends, entire from the middle downward, regularly deeply undulate upward, pinnately nerved; lower lateral nerves alternate, curving along the borders, camptodrome, mostly simple, the upper more oblique, simple or branching, reaching the borders at the outer end of the undulations, or broad round teeth.

The base of the lateral medial nerves is somewhat decurrent in joining the midrib at an acute angle of divergence, while the lower ones, more open, join it in a broad curve nearly at right angles, all more or less curving in passing to the borders. The leaf is 5 centimeters long without the short petiole (about 1 centimeter long), and nearly as broad. The only leaf known to me, to which this might be compared, is *Parrotia pristina*, Heer, "Fl. Arct.," vol. iv, p. 83. pl. xxi, fig. 5 (*Quercus fagifolia*, Goepp.), from which it differs not only by the leaf being shorter and broader, but by the distribution of the lateral nerves, the two lower pairs being alternate and at a short distance from each other, as in *Alnus serrulata*, Linn., while the upper, sub-opposite, parallel, and more distant, are branched and reach the borders at a more acute angle of divergence and a less pronounced curve.

Hab.—Four miles northeast of Minneapolis, Kansas. *Chs. Sternberg.*

Hamamelites quadrangularis, Lesqx.

Hayden's "Ann. Rep.," 1974, p. 355.
Alnites quadrangularis, Lesqx., "U. S. Geol. Rep.," vi, p. 62, pl. iv, fig. 1.

The leaf is small, slightly more coriaceous than the one described above; the borders are less distinctly undulate, and the secondary nerves thick, closely parallel, less divided; the two lower pairs of nerves are thinner and closer, following the borders like marginal nerves.

Hamamelites Kansaseanus, Lesqx.

Plate IV, Fig. 5.

Hayden's "Ann. Rep.," 1874, p. 355.
Alnus Kansaseanus, Lesqx., "U. S. Geol. Rep.," vi, p. 62, pl. xxx, fig. 8.

From the specimen figured here, which is better preserved than that copied in the "U. S. Geol. Rep." vi, the description is somewhat modified. The leaves are small, obovate in outline, cordate or obtuse at the gradually narrowed base: the borders are deeply regularly undulate from below the middle; the two lower pairs of lateral nerves thinner than those above and more open are camptodrome, the other craspedodrome. The basilar border seems to pass over the top of the petiole as in *Menispermites*.

Hab.—This species is not rare in Kansas. The specimen figured was communicated by Prof. *B. F. Mudge*. No. 698 of the National Museum.

Hamamelites quercifolius, sp. nov.

Leaf oblong, coriaceous, lanceolate, rounded to the base, blunt at the apex, undulate on the borders; nervation pinnate, deep; lateral veins close, oblique, craspedodrome, branching on the lower side.

The leaf has great likeness to *Dryopyllum (Quercus) latifolium*, pl. iv, fig. i. It is about the same length but narrower, only 5½ centimeters broad in the middle, as in the preceding species; the two lower pairs of secondary nerves are thinner, less oblique, more open than the eight others above. These slightly curve in passing to the borders and enter, like the divisions, the outside curve of the undulations.

Hab.—Bluff Creek, Ellsworth County, Kansas. *Chs. Sternberg.* There is only one specimen (No. 62*a* of the Museum Comp. Zool., Cambridge).

Hamamelites? cordatus, sp. nov.

Plate IV, Fig. 3.

Leaves large, thickish, broadly oval-oblong, deeply narrowly cordate at base, obtusely dentate; nervation pinnate; lateral nerves oblique, slightly curved in passing toward the borders, much branching on the lower side, craspedodrome.

This fragment represents a leaf about 12 centimeters long, 7 to 8 centimeters broad. It is undulate-dentate all around, pinnately nerved, with the secondary nerves at equal distance, and parallel, except two pairs of smaller ones attached to the base of the lower lateral nerves. Of these, the upper curves downward, branching and entering the borders by its apex and by its divisions, the lowest, simple and marginal, follows the nearly auricled basal borders. Nothing is seen of the areolation. Some simple parallel nervilles continuous and at right angles to the veins are seen in the upper part of the leaf, which by its facies and some of its characters resembles a *Viburnum*.

Hab.—Near Fort Harker, Kansas. *Chs. Sternberg.*

MAGNOLIACEÆ.

MAGNOLIA, Linn.

Magnolia alternans, Hoer.

"U. S. Geol. Rep.," vi, p. 93. pl. xviii, fig. 4.

Better specimens of this species, though not many, have recently been found in Kansas.

Magnolia Capellini, Heer.

"Phyll. Crét. du Neb.," p. 21, pl. iii, figs. 5, 6.

Leaves coriaceous, broadly oval, very entire; secondary veins at an acute angle of divergence, curving to the borders, camptodrome.

The leaves of this species are similar in size and shape to those described as *Ficus magnoliæfolia*, pl. xvii, figs. 5 and 6. This last figure, especially, does not differ from those published by Heer, except by the closer secondary veins and by the base, which is slightly decurrent in the leaves of *Ficus*, while in fig. 5 of Heer it is abruptly rounded and subcordate or subauricled. This appearance, however, may be merely casual, resulting from the breaking of the base, as seen in all the leaves of this species described by Heer in "Fl. Arct.," vol. iv, pl. xxxiii. Two specimens of this species found in Colorado have the base decurrent upon a short petiole, and the nervation of the species.

Hab.—The two specimens mentioned above (Nos. 12 and 12*b* of the collection of the Museum of Comp. Zool., of Cambridge) are from Morrison, Colorado, found by *A. Lakes*. I have received a number of others more or less fragmentary from Kansas.

Magnolia speciosa, Heer.

"Molet. Fl.," p. 20, pl. vi, fig. 1; ix, fig. 2; x; xi, fig. 1.

Leaves large, coriaceous, elliptical-ovate, narrowed upward into a long acumen and downward to a thick petiole; medial nerve thick; secondary nerves curved, camptodrome. (Heer.)

The leaves of this species are enlarged in the middle and more rapidly attenuated to a long acumen and to the petiole than in the preceding. The medial nerve is much thicker. The specimens which I refer to it differ in nothing from Heer's figure except, perhaps, by the lateral nerves, which appear somewhat closer. As the veins are very indistinct the reference is somewhat uncertain.

Hab.—Near Morrison, Colorado. *A. Lakes.* Specimen Nos. 13 and 13*a* of the Museum Comp. Zool., of Cambridge.

Magnolia tenuifolia, Losqx.

"U. S. Geol. Rep.," vi, p. 92, pl. xxi, fig. 1.

Magnolia obovata, Newby.

"Later Ext. Fl.," p. 15; "Illustr.," pl. ii, fig. 2; iv, fig. 4.

Leaves large, obovate, entire, thick and smooth, pointed and slightly decurrent on the petiole; nervation strong; midrib straight and extending to the summit; lateral nerves pinnate, set at somewhat unequal distances, straight and parallel below, forked and inosculating above, forming a festoon parallel with the margin; terminal nerves forming an irregular network of polygonal and relatively large areoles. (Newby.)

Hab.—Blackbird Hills, Nebraska. Dr. *Hayden.*

Magnolia species.

Plate XI, Fig. 6.

A flattened immature receptacle or carpile of a *Magnolia.* The short-pedicoled cone is oblong-obtuse, covered with short obtuse carpels.

Hab.—Near Morrison, Colorado. *H. C. Beckwith.*

LIRIODENDRON, Linn.

Liriodendron Meekii, Heer.

"Phyll. Crét. du Neb.," p. 21, pl. iv, figs. 3, 4.

Leaves panduriform, emarginate at the top, bilobate; lobes obtuse; secondary veins branching. (Heer.)

Hab.—Tekamah, Nebraska. Professor *Capellini.*

Liriodendron primævum, Newby.

"Later Ext. Fl.," p. 12; "Illustr.," pl. vi, figs. 6, 7.

Leaves three-lobed, upper lobe emarginate, all the lobes rounded; nervation delicate, medial nerve straight or slightly curved, terminating in the sinus of the superior lobe; secondary nerves gently arching upward, simple or forked near the extremities, a few more delicate ones alternating with the stronger. (Ny.)

From comparison of specimens received from Greenland, Professor Heer considers this species, as also the leaves described as *Leguminosites Marcouanus*, Heer, and *Phyllites obcordatus*, Heer (Newby., "Illustr.," pl. v, figs. 2, 3), as identical with the preceding species.

Liriodendron intermedium, Lesqx.

"U. S. Geol. Rep.," vi, p. 93, pl. xx, fig. 5.

No other specimen has been found as yet than the fragmentary one described in the "Report."

Liriodendron giganteum, Lesqx.

U. S. Geol. Rep.," vi, p. 93, pl. xxii, fig. 2.

A number of well-preserved specimens, recently obtained in Kansas, distinctly display the characters of this species originally described from a fragment, the upper lobe of a leaf only. The leaves are very large, 20 centimeters broad between the lower lobes, which are broad (6 centimeters), oblong, rounded or obtuse, at right angles to the medial nerve; upper lobes more oblique, shorter, narrowed and rounded to an obtuse point, joining the lower in a narrow deep sinus at a short distance (2 centimeters) from the thick medial nerves; lateral nerves parallel, nearly at equal distances, slightly oblique, curved down in joining the medial nerve.

By the form of the leaves this species is more than any other related to the living *L. Tulipifera*. As far as can be seen from the fragment of *L. intermedium*, this last species differs much from *L. giganteum*, especially by the deeply emarginate leaf, the very oblique upper lobes at a great distance from the lower ones. The facies of the leaves of these two species is far different.

Hab.—Two miles from Glasco, Kansas. The specimens, Nos. 206, 513, 535, found by *Chs. Sternberg*, like those of the four following species, belong to the collection of the Museum Comp. Zool., Cambridge.

Liriodendron acuminatum, Lesqx.

"Bull. Mus. Comp. Zool., Cambridge," vol. vii, No. 6, p. 227.

Leaves small, about half as large as those of the preceding species, cut into two pairs of narrow linear accuminate lobes all arched upward, about 10 to 12 centimeters long.

A remarkable species; the lobes, 1 centimeter broad, have only a medial nerve.

Hab.—Same location as the preceding. Specimens Nos. 476, 504, 504a.

Liriodendron cruciforme, Lesqx.

Ibid., p. 227.

Leaves large; upper lobes broad, square or equilateral, at right angles to the broad midrib; lower lobes narrow, linear, acuminate, much longer and arched upward.

The shape of the leaves is like that of an anchor, except that the medial nerve, or axis, does not pass above the upper border of the leaf, which is cut flat, not, or scarcely, emarginate.

Hab.—Elkhorn Creek. Nos. 197, 198, and some fragmentary ones.

Liriodendron semi-alatum, Lesqx.
"Bull. Mus. Comp. Zool., Cambridge," vol. vii, No. 6, p. 227.

Leaves divided at the base in two opposite short round lobes, obliquely cut in curving up to near the medial nerve and then diverging and enlarging upward into an obovate or spatulate entire lamina.

This form is somewhat like fig. 7 of pl. vi, Newby., "Illustr.," the lower lobes longer obtuse and more defined, the upper part gradually enlarged, spatulate, obtuse. It may be a distant form of *L. Meekii*.

Hab.—Seven miles from Glasco, Kansas. Specimens Nos. 472, 425.

Liriodendron pinnatifidum, Lesqx.
Ibid., p. 227.

A simple leaf, with the general facies and the nervation of *Liriodendron*, but narrow linear in outline, subalternately trilobed on each side. The top and base of the leaf are broken, the lobes, separated by broad flat sinuses, are half round, entire or irregularly undulate. The fragment is 9 centimeters long and 5 broad between the outside curves of the medial lobes, which are a little larger than the upper and lower ones; the lateral veins are close, oblique, parallel, distinct only at and near their point of union to the midrib. The fragment may represent a leaf of a different genus, though its affinity is evidently with *Liriodendron*.

Hab.—Two miles from Glasco, Kansas. Specimen No. 531 (526? fragment).

LIRIOPHYLLUM, Lesqx.

Leaves subcoriaceous, square or broadly rhomboidal in outline, abruptly narrowed to a comparatively short petiole, split from the top to the middle along the line of the medial nerve into two primary lobes much enlarged in the lower part, entire or sublobate or distinctly bilobate; nervation pinnate.

By the facies and the nervation these leaves have a great affinity to those of *Liriodendron*. Instead of being merely emarginate at the top they are deeply cut down, nearly to the middle, in two lobes joined by a narrow more or less obtuse sinus. This is indeed the more marked difference.

Liriophyllum Beckwithii, Lesqx.

Plate X. Fig. 1.

Hayden's "Ann. Rep.," 1876, p. 4~2, mentioned.

Leaves large, square in outline, cut to near the base into two large diverging lobes; lobes bilobate, obtuse; primary nerve very thick, continuous to a short petiole, bifid at a short distance above the base, the divisions ascending to the obtuse point of the upper lobes; secondary veins two, parallel, curved into the lower lobe, all with few branches.

The abnormal form of the leaves of this genus renders their description difficult. In this species, which may be a variety or deformation of the following, the leaves are large, about 28 centimeters between the points of the lower lobes, and nearly 20 centimeters from the base to the apex of the upper. They are divided into two halves from the top to 4 centimeters above the base by the splitting of the medial nerve under an angle of 40°, and each division is cut at the side in two short obtuse lobes separated by a broad sinus. The lower lobe, nearly at right angles to the midrib, is traversed in its whole length by two parallel, strong, secondary nerves, apparently vanishing below the top (broken). Except very few oblique curved tertiary veins, no other trace of nervation or areolation is distinct. The medial nerve from under the sinus downward is 3 millimeters broad—as broad as the short pedicel broken 2 centimeters below the slightly decurrent base of the leaf.

Hab.—Near Morrison, Colorado. *H. C. Beckwith.* Found only in one good specimen.

Liriophyllum populoides, Lesqx.

Plate XI, Figs. 1 and 2.

Leaves smaller, broadly ovate, cuneiform at base, divided nearly vertically from the top to above the middle into two obtuse lobes, enlarged on the rounded sides above the base and there sometimes prolonged into a short obtuse lobe; medial nerve straight; lateral nerves strong, parallel, equidistant, four pairs, effaced near the borders, rarely branching; nervilles at right angles.

In comparing these leaves with the preceding the essential characters are seen to be identical, though the appearance is far different. The large size of the leaf and the subdivision of the two primary lobes in *L. Beckwithii* are the more marked differences. But in fig. 1 of this species the lower side is continued into a short lobe, indicating a subdivision like that of the leaf pl. x, fig. 1, and the nervation is of the same type as in the leaf pl. xi, fig. 1; the two lower lateral nerves turn outside toward the short lobes, while the upper is evidently tending upward.

Hab.—With the preceding in numerous specimens. *A. C. Beckwith, A. Lakes.* One specimen also has been found in Kansas.

Liriophyllum obcordatum, sp. nov.

Leaf small, obovate, entire, narrowly deeply emarginate at the top, gradually narrowing to the petiole (broken); medial nerve narrow; lateral nerves at an acute angle of divergence, alternate, camptodrome.

This leaf, 6 centimeters long and 3 broad above the middle, is cut from the top to one-third of its length into two obtuse entire slightly diverging lobes by the splitting of the medial nerve, as in the two preceding species. It is perfectly entire, gradually narrowed from above the middle, or cuneiform to the base, with two pairs of alternate distant secondary nerves at an acute angle of divergence and curving in passing toward the borders. The tertiary nervation and the areolation are totally obsolete.

Hab.—With the preceding. Rev. *A. Lakes.*

Carpites liriophylli? sp. nov.

Plate XI, Fig. 5.

An oblong seed 3 centimeters long. 7 millimeters broad in the middle, narrowed and blunt at one end, acute at the other; irregularly obscurely lineate on the surface.

The reference of this fruit to *Liriophyllum* is hypothetical. The seed was found on one of the specimens of *M. Beckwith*, with leaves of *L. populoides*.

ANONACEÆ.

ANONA, Linn.

Anona cretacea, sp. nov.

Leaf lanceolate or oblong-lanceolate, gradually narrowed to a short flattened petiole; medial nerve thick; secondary nerves open, nearly at right angles toward the base, branching, camptodrome.

A fragment of leaf of which the lower half only is well preserved. It is similar in its size, form, and venation to *A. lignitum.* Ung., "Syllog.,"

p. 25, pl. x, figs. 1–6. The relation of this leaf to this genus is as evident as it can be indicated by a single specimen representing only part of a leaf and no fruit.

Hab.—Near Glasco, Kansas. *Chs. Sternberg.* No. 414 of the collection of the Museum of Comp. Zool., Cambridge.

MENISPERMACEÆ.

MENISPERMITES, Lesqx.

"U. S. Geol. Rep.," vi, p. 94.

The definition of this genus has to be somewhat modified in this: the leaves are not only broadly deltoid- and more or less distinctly trilobate, but also round or ovate, entire, with a camptodrome nervation. From this, the group is subdivided in two sections, represented one by lobate, the other by entire leaves.

Menispermites obtusilobus, Lesqx.

Plate XV, Fig. 4.

"U. S. Geol. Rep.," vi, p. 94, pl. xxv, figs. 1, 2; xxvi, fig. 3.
M. obtusilobus var., ibid, p. 95, pl. xxii, fig. 1.

Menispermites Salinensis, Lesqx.

"U. S. Geol. Rep.," vi, p. 95, pl. xx, figs. 2, 3.

Menispermites acutilobus, sp. nov.

Plate XIV, Fig. 2.

Leaf large, triangular in outline, broadly rounded or nearly truncate at base, deltoid, dentate-lobate, five-nerved from near the base, coriaceous; nerves more or less branching on the lower side, craspedodrome, with their divisions; nervilles at right angles to the nerves, anastomosing in the middle of the areas.

The specimen figured is the only one seen. Comparison of the figures representing this species and *M. obtusilobus*, pl. xv, fig. 4, shows the close affinity of the leaves—*M. acutilobus* merely differing by the large acute distant teeth of the borders. The primary nervation is the same as that in pl. xv, fig. 1; the secondary veins are distant, equally oblique, and curving toward the borders, scarcely branching, all craspedodrome, and entering the teeth of the borders, a character already remarked in all the specimens of *M. obtusilobus*, whose secondary veins are more generally craspedodrome even when the borders are not undulate-dentate, and always so when the leaves are undulate.

Hab.—Clay County, Kansas. *H. C. Towner.*

Menispermites populifolius, Lesqx.

Plate IV. Fig. 4.

Hayden's "Ann. Rep.," 1874. p. 357.

Leaf broadly ovate, obtuse, subcordate or truncate at base, palmately five-nerved from near the basal borders; primary lateral nerves at a more acute angle of divergence, branching on the lower side; secondary nerves equidistant, parallel, all camptodrome.

The leaf is coriaceous, smooth on the surface, perfectly entire, 5½ centimeters long and as broad in its largest diameter below the middle. The primary lateral veins diverge about equally from each other at an angle of about 30°; the lower is nearly simple and has still a thin marginal veinlet underneath; they branch from the lower part, and the secondary nerves at a distance above fork only at their ends toward the borders. The areas are crossed by very strong nervilles at right angles to the nerves, anastomosing in the middle. The areolation is obsolete.

Hab.—South of Fort Harker. *Chs. Sternberg.*

Menispermites cyclophyllus, Lesqx.

Plate XV, Fig. 3.

Hayden's "Ann. Rep.," 1874, p. 358, pl. vi, fig. 4.

Leaf thick, subcoriaceous, very entire, nearly round and centrally peltate, deeply concave, palmately five-nerved; inner lateral nerves curving inside, the outer open, nearly at right angles to the medial nerve, all dividing by open straight branches anastomosing at a distance from the borders in double rows of arches; basilar veins 3 to 5, diverging star-like from the central point.

The leaf is 7 centimeters long and 6 broad in its widest diameter; the middle is rounded downward and a little more narrowed upward to the round subtruncate apex. The point of attachment of the petiole is nearly central, and though surrounded by a series of nerves diverging star-like, it has, like the other species of this genus, five primary nerves turning upward, the lower ones representing marginal veins. The leaf is concave from the point of attachment of the petiole, which passes down into the stone, leaving an opening like the pipe of a funnel.

Hab.—Near Fort Harker, Kansas. *Chs. Sternberg.*

Menispermites grandis, sp. nov.

Plate XV, Figs, 1, 2.

Leaves subcoriaceous, large, flat, nearly round, broader than long, peltate; borders entire or undulate; nerves radiating from the point of attachment, camptodrome; primary nerves five.

This species differs from the preceding not only by the large size of the leaves but especially by the nervation which is simply camptodrome, the veins and their divisions curving along close to the borders and anastomosing in a single row of festoons. Even the medial nerve has the same character and does not ascend to the borders, but is forked near the apex in camptodrome divisions.

Hab.—Near Clay Centre, Kansas. *H. C. Towner.*

Menispermites ovalis, Lesqx.

Plate XV. Fig. 5.

Hayden's "Ann. Rep.," 1874. p. 357, pl. v, fig. 4.

Leaf narrowly oval or oblong, obtusely pointed, rounded at base, palmately fivenerved; lateral nerves at an acute angle of divergence, the inner ones ascending to near the top, branching outside; branches numerous, parallel, curving along the borders in festoons.

This fine leaf, preserved nearly entire, is 7 to 8 centimeters long, 3½ centimeters broad, nearly exactly oval-oblong, perfectly entire. It is less distinctly palmately five-nerved than the leaves of the other species of this genus; the two internal primary nerves are as strong as the medial one, curve gradually nearly parallel to the borders, and near the top join the branches of the midrib with which they anastomose in curves; the outside lateral nerves are thinner and shorter; they ascend also nearly parallel to the borders, disappearing in the middle of the leaf in anastomosing with branches of the lateral primary nerves. This is a mere deviation from the type.

Under the name of *Daphnogene Kanii*, Heer has published ("Fl. Arct.," i, p. 112, pl. xiv), from the Miocene of Greenland, leaves related by their form to this Cretaceous species. The same kind of leaves are described by Saporta and Marion in the "Flora of Gelinden," p. 63, pl. x, as *Cocculus Kanii.* In these leaves the primary nervation is in three from the base; in the Cretaceous leaf it is positively in five and therefore different, appearing intermediate between that of the leaves described above as *Menispermites* and that of *Daphnogene*, or *Cocculus Kanii.*

Hab.—Near Clay Centre, Kansas.

MALVACEÆ.

STERCULIA, Linn.

Leaves alternate, petiolate, palmately deeply trilobate; triple-nerved from the top of the petiole.

This definition represents the characters of the coriaceous leaves which I refer to this genus, and which I separate from *Aralia* merely on account of the primary divisions. Most of these leaves have only the primary nerves distinct and rarely any trace of the secondary veins. By a lower division of the lateral primary nerves, species referable, perhaps, to this genus are described above as *Aralia*. If, as Schimper says, *Sterculia Majoliana*, Massal., "Fl. Foss. Senig., p. 319," is referable to the group of *Sterculia Labrusca*, most of the species that are described as *Aralia*, if not all, should be placed also with *Sterculia*. I do not admit this conclusion.

Sterculia lugubris, sp. nov.

Plate VI, Figs. 1-3.

Leaves coriaceous, large, divided near the cuneate base into three very long sub-linear acuminate lobes; primary nerves thick, distinct to the apex.

The leaves, narrowly cuneate, somewhat decurrent at base to the thick petiole, which they reach a little below the point of union of the primary nerves, vary in length from 12 to 24 centimeters from the base to the apex of the lobes, which are united by obtuse comparatively narrow sinuses at a short distance—3 to 6 centimeters—from the base. The lobes, 1 to 2 centimeters broad in the middle, are slightly narrowed to their base, and gradually tapering from the middle upward to an acuminate point. The lateral are curved downward, or scythe-shaped. No trace of secondary nervation is visible.

There is in the collection of the National Museum a set of specimens representing an analogous form, though perhaps specifically different. The lobes, descending nearer to the base, are shorter (7–14 centimeters long), straight, not recurved, linear-oblong, slightly narrowed from the middle downward to the broad obtuse sinuses and gradually to the apex. All the points of the lobes are destroyed. Their divergence is about 25°.

Hab.—Colorado, near Golden. *A. Lakes.* The variety is from Kansas. *Chs. Sternberg.*

c f G

Sterculia obtusiloba, Lesqx.

Plate VIII, Fig. 3.

Aralia tripartita, Lesqx., Hayden's " Ann. Rep.," 1874, p. 348, pl. i, fig. 1.

Leaves coriaceous, small, palmately three-lobed; lobes equal, linear, obtuse, very entire; secondary nerves obsolete.

The only leaf I have seen of this species is figured. It is 7 centimeters long, 6 centimeters broad between the points of the lateral lobes, which diverge at an angle of 25° and are cut down to about two-thirds of the leaf. The medial lobe is a little narrower than the lateral (1 centimeter broad); the leaf is cuneate to the base and apparently a little decurrent to the petiole (broken); its surface is smooth. This leaf, following the definition of the genus, represents a *Sterculia.* Its name was changed accordingly.

Hab.—Near Fort Harker, Kansas. *Chs. Sternberg.*

Sterculia aperta, sp. nov.

Plate X, Figs. 2, 3.

Leaves subcoriaceous, palmately three-lobed, and triple-nerved from near the base; lobes lanceolate, blunt at the apex; angle of divergence broad.

This species is different from the preceding by the form of the broader lanceolate obtusely pointed lobes, the leaves not as thick and larger. Fig. 3 shows traces of secondary nerves equidistant and curving to the borders, the lower ones on the medial nerve being at right angles to it. These leaves are related to *Sterculia labrusca,* Ung., a species which, already present in the Eocene of France, is found also in all the stages of the Tertiary, including the Pliocene, in very variable forms. A number of specimens in the Museum of Comp. Zool. of Cambridge represent a form which seems intermediate between this and the preceding. The leaves are 8 to 10 centimeters long, somewhat thick but not coriaceous, with lobes more or less diverging, linear-lanceolate, gradually narrowed above to a blunt point, nearly equal in length, 4 to 5½ centimeters long, 12 to 14 millimeters broad.

Hab.—Kansas. Found at divers localities. *Chs. Sternberg.*

TILIACEÆ.

GREVIOPSIS, Sap.

The remark made on the definition of this genus, "U. S. Geol. Rep.," vii. p. 257, is applicable also to the Cretaceous leaves which I have described under this generic name. The character of the nervation especially relates them to those figured by the celebrated author in the "Sézanne Flora."

Greviopsis Haydenii, Lesqx.

"U. S. Geol. Rep.," vi, p. 97, pl. iii, figs. 2, 4; xxiv, fig. 3.

The leaf represented in this last figure was described first in "Amer. Jour. Sci. and Arts," July, 1868, as *Populites flabellata*.

ACERACEÆ.

ACERITES, Newby.

Acerites pristinus, Newby.

"Later Ext. Fl.," p. 15; "Illustr.." pl. v, fig. 4.

Leaves petiolate, cordate at the base, five-lobed; lobes entire, acute;? five strong and nearly equal veins radiate from the base into the lobes. The small nerves are distributed over the surface in a fine net-work of which the meshes are sub-rectangular. (Nv.)

The figure represents a fragmentary leaf of the same character as those described and figured in "U. S. Geol. Rep.," vi, p. 56, pl. ii, figs. 1, 3, under the name of *Liquidambar integrifolium*. The relationship of these leaves seems to be with the *Araliaceæ*, but it is as yet unascertained.

Negundoides acutifolius, Lesqx.

"U. S. Geol. Rep.," vi, p. 97, pl. xxi, fig 5.

SAPINDACEÆ.

SAPINDUS, Linn.

Sapindus Morrisoni, sp. nov.

Plate XVI, Figs. 1, 2.

Leaflets subcoriaceous, short petioled, lanceolate-acuminate, unequal at the rounded narrowed slightly decurring base; lateral nerves alternate, parallel, curving in passing to the borders, camptodrome.

The fragment represents apparently the base of a large pinnately divided leaf, with leaflets alternate, short petioled, more **enlarged on one** side near the base. The fragments of leaflets distributed on the same piece of coarse shaly sandstone indicate their original connection with a pinnate leaf. The lower part of the stem does not bear any fragments of the base of other leaflets attached to it. The stone is coarse, the nervation is obscure and has no trace of subdivisions of the secondary veins. The leaflets average 12 to 14 centimeters in length, 2½ to 3 centimeters in width in the broadest part below the middle.

Hab.—Near Morrison, Colorado. *H. C. Beckwith.*

Fragments of what I consider a variety of this species have been sent by Chs. Sternberg to the Museum of Comp. Zool., Cambridge, from Ellsworth County, Kansas (*Nos.* 24, 37). These represent two leaflets only, both unequal at base, one about the same size as the specimens from Morrison, merely differing by the lateral veins being a little more oblique; another leaflet is shorter and has the veins open proximate. It has been found also at Atane with *S. prodromus,* Heer, "Fl. Arct.," iii, p. 117, pl. xxxiv, which it resembles.

FRANGULACEÆ.

CELASTROPHYLLUM, Ett.

Celastrophyllum ensifolium, Lesqx.

"U. S. Geol. Rep.," vi, p. 108, pl. xxi, figs. 2, 3.

ILEX, Linn.

Ilex strangulata, Lesqx.

Plate III, Fig. 7.

Hayden's "Ann. Rep.," 1874, p. 359, pl. vii, fig. 8.

Leaf coriaceous, narrow, panduriform or strangled in the middle to a small angular lobe, rounded at base in narrowing to the petiole, entire in the lower part, little enlarged and irregularly distinctly obtusely dentate in the upper; secondary veins proximate, in a very open angle of divergence, irregularly camptodrome or mixed.

This leaf is about 5½ centimeters long (point broken) without the 1½ centimeter long petiole. The general outline of the leaf is lanceolate, but it is narrowed in the middle, as by erosion, nearly to the medial nerve, and gradually enlarged upward by undulations or successive large obtuse irregular teeth. The surface is rugose; the lateral nerves, mostly camptodrome, follow close to the borders, the lower pair at a more acute angle of divergence as marginal veins, and those of the middle abruptly curved, following also close to the borders with the same appearance as that of the basilar nerves. This nervation is related to that of some species of *Myrica*, and still more of *Ilex*, like *I. Abichi, I. berberidifolia*, Heer, of the Miocene. The areolation, distinct only on a small area where the epidermis is destroyed, is in small, angular or irregularly square areoles. The narrowing of the leaf in the middle appears as produced by the gnawing of insects. But if the vein which follows the border is not a deceptive representation caused by the thickness of the leaf, this peculiar deformation is natural. Leaves of *Ilex* are often variously and abnormally cut.

Hab.—Same as *Dryophyllum (Quercus) Holmesii*, in connection with coal strata of Southwest Colorado at a higher stage of the Cretaceous. *H. Holmes.*

FRANGULACEÆ.

PALIURUS, Tourn.

Paliurus membranaceus, Lesqx.

" U. S. Geol. Rep.," vi, p. 108, pl. xx, fig. 6.

RHAMNUS, Juss.

Rhamnus tenax, Lesqx.

" U. S. Geol. Rep.," vi, p. 109, pl. xxi, fig. 4.

Rhamnus prunifolius, sp. nov.

Leaf coriaceous, ovate-lanceolate, rounded in narrowing to the base; medial nerve deep, straight; lateral nerves at short distance, parallel, open, arched in passing toward the borders and curving along and close to them; nervilles close, numerous, oblique to the nerves.

This leaf, 4 to 5 centimeters long (point broken), nearly 3 centimeters in the middle, resembles what Heer describes as *Salix nervillosa*, "Phyll. Crét. du Neb.," pl. i, fig. 3; but the lateral nerves are open, joining the medial nerve nearly at right angles, parallel from the base of the leaf, which is not cuneiform but more rounded; the nervilles are oblique to the veins. The nervation is that of a *Rhamnus*.

Hab.—Near Glasco, Kansas. *Chs. Sternberg.* No. 479 of the Museum Comp. Zool., Cambridge.

JUGLANDEÆ.

JUGLANS, Linn.

Juglans? Debeyana, Heer.

"U. S. Geol. Rep.," vi, p. 110, pl. xxiii, figs. 1–5.
Populus Debeyana, Heer, "Phyll. Crét. du Neb.," p. 14, pl. i, fig. 1; Newby., "Notes on Ext. Fl.," p. 17;
 "Illustr.," pl. iv, fig. 3.

ANACARDIACEÆ.

Phyllites rhoifolius, Lesqx.

"U. S. Geol. Rep.," vi, p. 111, pl. xxii, figs. 5, 6.

POMACEÆ.

PYRUS, Lindl.

Pyrus? cretacea, Newby.

"Notes on Ext. Fl.," p. 42; "Illustr.," pl. ii, fig. 7.

Leaves petioled, small, roundish, oval or elliptical, often slightly emarginate, entire or finely serrate; medial nerve strong below, rapidly diminishing toward the summit; lateral nerves four or five pairs, with intermediate smaller ones, diverging from the midrib at unequal angles, curved toward the summit, where they anastomose in a series of arches parallel with the margins; tertiary nerves forming a net-work of which the areoles are somewhat elongated. (Ny.)

This leaf seems to be a small, lateral leaflet of *Juglans? Debeyana.*

Hab.—Smoky Hills, Kansas. Dr. *Hayden.*

AMYGDALEÆ.

PRUNUS, Tourn.

Prunus cretacea, Lesqx.

"U. S. Geol. Rep.," vi, p. 111, pl. xxiii, figs. 8, 9.

LEGUMINOSÆ.

LEGUMINOSITES, Auct.

Leguminosites cultriformis, sp. nov.

Plate X, Fig. 4.

Fruit (legume) stipitate, rounded to the point of support, enlarged above it and gradually tapering up to an obtuse point; stipe enlarged at base.

The legume is 7½ centimeters long without its stipe (a little more than 2 centimeters), 13 millimeters broad above the base, the widest part, and gradually narrowed, by the inclination of one of its sides only, to a blunt point. The whole surface is smooth with only some fragments of longitudinal lines.

No remains of *Leguminosæ* have been discovered in the Dakota Group except the one figured as above. It appears to be a stipitate legume with analogy of form and size to those of *Lonchocarpus*, H. B. & Kunth., a genus mostly represented in the West Indian Islands, the equatorial America.

GENERA AND SPECIES OF UNCERTAIN RELATION.

ASPIDIOPHYLLUM, Lesqx.

Hayden's "Ann. Rep.," 1874, p. 361.

Leaves large, triangular in outline, palmately trilobate, truncate or rounded to a peltate base; nervation coarse; primary nerves trifid, from a short distance above the peltate base of the leaves, the lateral, at an open angle of divergence, sometimes curved downward; secondary nerves generally close, parallel, camptodrome, generally simple, joined by strong nervilles at right angles.

This group has a great affinity by the form of the leaves and the nervation to that of the *Sassafras* (Araliopsis). Indeed at first sight it appears to differ from it only by the addition of a basilar shield. The nervation, however, differs in some characters, the primary nerves being at a more open angle of divergence, as are also the secondary ones, which are also more curved in passing to the borders. The rounded more or less enlarged shield of the base is nerved by the secondary nerves gradually declining downward, one pair generally attached under the point of union of the primary nerves, the others derived from the base of the medial nerve and passing downward, the lowest nearly perpendicular in direction, and all abruptly curving and following the borders in continuous flexures. The disposition of the lower lateral nerves has an analogy to that of *Credneria*, with the difference that in *Credneria* the lower secondary nerves are all at right angles to the midrib. The same kind and degree of analogy is marked between these leaves and those of *Protophyllum* and *Pterospermites*, and also those of *Platanus*.

Aspidiophyllum trilobatum, Lesqx.

Plate XII, Fig. 1; XIII, Figs. 1–5; XIV, Fig. 1.

Leaves generally large, coriaceous, triangular or rhomboidal in outline, deeply obtusely trilobate, broadly cuneate to the base, enlarged into a half-round entire auricle. The leaves vary in size from 10 to 24 centimeters long and from 10 to 30 centimeters broad between the lateral lobes. Some of them, apparently constituting a variety of the normal form, are not half as large, their nervation is still coarser and the surface rugose, as in pl. xiii, fig. 1, and especially pl. xiv, fig. 1. All have been found at the same locality, mostly alone. There is also a marked difference in the expansion of the peltate base, which is generally half-round, as in pl. xiii, figs. 1, 3, but which sometimes is regularly dentate lobate around, as in pl. xiii, fig. 5. But this fragment may be referable to the following species.

Hab.—Found in numerous specimens 3½ miles south of Fort Harker. *Chs. Sternberg.*

Aspidiophyllum dentatum, sp. nov.

Leaves smaller, palmately three-lobate, peltate at the base; lateral lobes trilobate, the medial long, all dentate in the upper part; secondary nerves camptodrome; base of the leaves contracted into a fan-like five-lobed basilar shield.

The leaves have the same general facies as those of *A. trilobatum,* differing by their texture not being as thick, the nervation not as coarse, and by the base of the leaves being contracted under the point of division of the primary nerves into a narrow neck half a centimeter broad only, and then abruptly enlarged into a fan-like five-lobed or deeply dentate shield or stipule 4 centimeters broad between the summits of the lateral teeth and 2 centimeters vertically from the base of the medial nerve to the end of the lower lobes. This form or species with the dentate borders of the middle lobes and the subdivisions of the lateral lobes has its affinity to *Sassafras (Araliopsis) cretaceum,* while the preceding species has it to *S. (Araliopsis) mirabile.* Another specimen of the same group shows the basilar shield transversely oval and entire, stipuliform, also separated from the leaf by a narrow neck. But of this I have seen only a mere fragment, the base of a leaf. It possibly represents still another species.

Hab.—Eight miles northeast of Minneapolis. *Chs. Sternberg.* Specimens 607 and 614 of the Museum Comp. Zool. Cambridge.

Aspidiophyllum platanifolium, sp. nov.

Plate II, Fig. 4.

Leaves of various sizes, thinner or not coriaceous, rhomboidal in outline, irregularly short trilobate, triple-nerved high above the base; secondary nerves distant and irregular in position and direction, craspedodrome, with camptodrome divisions.

The few leaves I have seen of this species are about of the same size,

15 centimeters long, 13 centimeters broad between the lateral short broadly obtuse lobes. The substance of the leaves is not coriaceous, rather thin or membranaceous; the nervation not as coarse; the primary veins only half as thick as in the preceding species. The medial nerve descends to near the basilar margin before passing under it, and thus the tertiary or marginal veins join the lower part of the medial nerve at right angles as in *Credneria;* the upper secondary nerves, only three pairs, are very distant and oblique, not parallel nor equal in distance, and reach the margins by their ends as craspedodrome while all their divisions are camptodrome. The relation of this leaf to *Platanus* is quite distinct, as will be seen in comparing it to *P. Heerii,* "U. S. Geol. Rep.," vi, pl. ix, fig. 1.

Hab.—Clay County, Kansas. *H. C. Towner.*

Protophyllum, Lesqx.
" U. S. Geol. Rep.," vi, p. 100.

Protophyllum Sternbergii, Lesqx.
Ibid., p. 101, pls. xvi, xviii, fig. 2.

Protophyllum Leeonteanum, Lesqx.
Ibid., p. 103, pl. xvii, fig. 4; xxvi, fig. 1.

Protophyllum Nebrascense, Lesqx.
Ibid., p. 103, pl. xxvii, fig. 3.

Protophyllum quadratum, Lesqx.
Ibid., p. 104, pl. xix, fig. 1.

Protophyllum minus, Lesqx.
Plate IV, Fig. 6.
Ibid., p. 104, pl. xix, fig. 2; xxvii, fig. 1.

This species sometimes has the leaves very rugose and thus resembles *P. rugosum,* which is, however, very different in the nervation, the large size of the leaves, etc.

Protophyllum multinerve, Lesqx.
Ibid., p. 105, pl. xviii, fig. 1.

From numerous specimens less fragmentary than the one figured the leaves are seen to be round or transversely oval with borders entire. The sizes vary from 7 to 14 centimeters long and 9 to 18 centimeters broad. The nerves are very close and numerous around the peltate base of the leaves; above it they count 8 to 10 pairs, the lower forking generally once, the upper simple.

Protophyllum rugosum, Lesqx.

"U. S. Geol. Rep.," vi, p. 105, pl. xvii, figs. 1, 2; pl. xix, fig. 3.

Among other leaves of this species there is one entirely preserved, No. 747, in the Museum Comp. Zool. Cambridge. It measures 17 centimeters long, 10 broad, is undulate on the borders or somewhat dentate by the projection of the lateral veins, and agrees in every point by form and nervation with the figure and description of the species (loc. cit.).

Protophyllum Haydenii, Lesqx.

Ibid., p. 106, pl. xvii, fig. 3.

Protophyllum crednerioides, Lesqx.

Plate II, Figs. 1–3.

Hayden's "Ann. Rep.," 1874, p. 363, pl. iii, fig. 1; viii, fig. 4.

Leaves small, nearly round, broadly cuneate or subtruncate at base, long-petioled; borders entire or more generally undulate; nervation obscurely trifid; secondary veins parallel, equidistant, at various angles of divergence, more or less branching.

The leaves vary in size from 6 to 8 centimeters both ways. The borders are either deeply undulate or nearly entire, though all the nerves and their divisions are craspedodrome; the secondary nerves are open, at right angles toward the base. The areolation is formed by anastomosing of continuous nervilles at right angles to the veins and by their subdivisions in the areas, also at right angles, forming very small quadrate meshes, as seen in fig. 3. As in the other species of the genus, the nervation is more or less obscurely trifid. The lower primary lateral nerves being at a distance above the borders have under them, as in Credneria, two pairs of thinner secondary or marginal nerves at right angles. But as the lower veins often branch like the upper ones and have the same direction as those above, the nervation sometimes appears pinnate, as in fig. 1. The ternate disposition is, however, distinct in fig. 3.

Hab.—Kansas. Not rare. Chs. Sternberg, H. C. Towner.

Protophyllum? Mudgei, Lesqx.

"U. S Geol. Rep.," vi, p. 106, pl. xviii, fig. 3.

ANISOPHYLLUM. Lesqx.

"U. S. Geol. Rep.," vi, p. 98.

Anisophyllum semi-alatum, Lesqx.

Ibid., p. 98, pl. vi, figs. 1-5.

No other specimens have been seen of this species since it was first examined.

EREMOPHYLLUM, Lesqx.

Ibid., p. 107.

Eremophyllum fimbriatum, Lesqx.

Ibid., p. 107, pl. viii, fig. 1.

The specimen figured is the only one seen of this kind.

VEGETABLE REMAINS OF UNCERTAIN AFFINITY.

PHYLLITES, Auct.

Phyllites Vanonae, Heer.

Ibid., p. 113, pl. xx, fig. 7 ; xxviii, fig. 8.

Phyllites rhomboideus, Lesqx.

Ibid., p. 112, pl. vi, fig. 8.

Phyllites cotinus, Lesqx

Hayden's " Ann. Rep.," 1874, p. 364.
Bumelia Marcouana, Heer, " U. S. Geol. Rep.," vi, p. 90, pl. xxviii, fig. 2.

Phyllites umbonatus, Lesqx.

"U. S. Geol. Rep.," vi, p. 113, pl. xix, fig. 4.

Phyllites amorphus, Lesqx.

Ibid., p. 113, pl. xxii, figs. 3, 4.

Ptenostrobus, Lesqx.

Ibid., p. 114.

Ptenostrobus Nebrascensis, Lesqx.

Ibid., p. 114, pl. xxiv, fig. 1.

CARPOLITHES, Auct.

Carpolithes species?

Ibid., p. 114, pl. xxvii, fig. 5 ; xxx, fig. 11.

CAUDEX.

Caudex spinosus, Lesqx.

Caulinites spinosus, Lesqx., *Ibid.,* p. 115, pl. xxvii, fig. 4.

CONCLUDING REMARKS.

The Flora of the Dakota Group, as already remarked, is considered as relating the formation which it represents to the Cenomanian or Middle Cretaceous. In order to ascertain the validity of the relationship, and also to have a clear exposition of the general characters of the vegetation of the time, I have prepared the following table of the species of fossil plants which have been described by authors as referable to that stage of the Cretaceous.

1st. Those from Atane, Greenland; described by Heer in the "Fl. Arct.," including part ii of vol. vi, recently published.

2d. The species known from Molelein and Quedlinburg, described by the same author.

3d. The plants found in the Quader sandstone of the Hartz and of Bohemia, described or mentioned in different memoirs by Hampe, Stiehler, Dunker, Goeppert, Feistmantel, Corda, etc.

4th. The species described from Niedershœna in Saxony, by d'Ettingshausen.

TABLE OF DISTRIBUTION OF THE PLANTS OF THE CRETACEOUS CENOMANIAN FORMATION.

Number.	NAMES OF SPECIES.	Dakota Group—Kansas, Nebraska, Minnesota.	Dakota Group—Colorado, Eastern base of the Mountains.	Greenland—Schists of Atane.	Europe—Moletein, Quedlinburg.	Quader Sandstone—Hartz, Bohemia.	Niederschena—Saxony, Hungary.	Lower Cretaceous.
	CRYPTOGAMÆ.[1]							
	Equisetæ.							
1	Equisetum nodosum, Lx	+						
2	Equisetum amissum, Hr				+			+
	Filices.							
3	Sphenopteris corrugata, Ny	+						
4	Hymenophyllum cretaceum, Lx	+						
5	Cynthea fertilis, Hr			+				
6	Cyathea Hammeri, Hr			+				
7	Dicksonia Groenlandica, Hr			+				
8	Dicksonia borealis, Hr			+				
9	Dicksonia conferta, Hr			+				
10	Dicksonia punctata, Hr			+				
11	Pteris frigida, Hr			+				+
12	Pteris longipennis, Hr			+				
13	Pteris ? Albertsii, Dkr			+				
14	Pecopteris lobifolia, Corda					+	+	+
15	Pecopteris arctica, Hr			+				
16	Pecopteris striata, St					+	+	
17	Pecopteris argutula, St					+	+	
18	Pecopteris linearis, St			+			+	+
19	Pecopteris borealis, Brgt			+				
20	Pecopteris Pfaffiana, Hr			+				
21	Pecopteris Reichiana, Ett						+	
22	Pecopteris denticulata, Hr			+				
23	Pecopteris Bohemica, Corda			+		+		
24	Pecopteris Nebraskana, Hr	+						
25	Pecopteris socialis, Hr			+				
26	Aspidium Reichianum, St						+	
27	Aspidium Œrstedi, Hr			+				
28	Aspidium fecundum, Hr			+				
29	Aspidium Schouwii, Hr			+				
30	Aspidium Jenseni, Hr			+				
31	Phegopteris Jörgenseni, Hr			+				
32	Asplenium Dicksonianum, Hr			+			+	
33	Asplenium Fosteri, Deb			+				
34	Asplenium Nordstromi, Hr			+				
35	Gleichenia Giesekiana, Hr			+		+		+
36	Gleichenia gracilis, Hr			+				+
37	Gleichenia acutiloba, Hr			+				
38	Gleichenia rigida, Hr			+		+		
39	Gleichenia Nauckhoffii, Hr			+				
40	Gleichenia Kurriana, Hr	+			+	+		+
41	Gleichenia Zippei, Hr			+	+	+		+
42	Gleichenia Nordenskiöldi, Hr	+						+

[1] I omit in this table the Thallophytes represented by six species of fungi upon leaves of Niederschena and of Atane, and the marine Zonarites found in connection with animal remains, especially mollusks, in strata overlying the Dakota Group.

Table of Distribution of the Plants of the Cretaceous Cenomanian Formation—Continued.

Number	NAMES OF SPECIES.	Dakota Group—Kansas, Nebraska, Minnesota.	Dakota Group—Colorado, Eastern base of the Mountains.	Greenland—Schlais of Atane.	Europe—Moletein, Quedlinburg.	Quader Sandstein—Hartz, Bohemia.	Niedersachsen—Saxony, Hungary.	Lower Cretaceous.
43	Gleichenia comptoniæfolia, Ett.			+		+	+	+
44	Gleichenia obtusata, Hr.			+				
45	Lygodium trichomanoides, Lx.	+						
46	Osmunda Obergiana, Hr.			+				
47	Weichselia Ludovicæ, Stiehl.				+	+		
	Rhizocarpeæ.							
48	Marsilea cretacea, Hr.			+				
	Selagineæ.							
49	Selaginella arctica, Hr.			+				
	PHENOGAMEÆ.							
	Cycadeæ.							
50	Cycas Steenstrupi, Hr.			+				
51	Cycadites Dicksoni, Hr.			+				
52	Pterophyllum cretosum, Reich.						+	
53	Pterophyllum Saxonicum, Reich.					+	+	
54	Zamites latipennis, Hr.			+				
55	Podozamites Haydenii, Lx.	+						
56	Podozamites Ernestinæ, Stieh.					+		
57	Podozamites marginatus, Hr.			+				
58	Podozamites minor, Hr.			+				
59	Podozamites tenninervis, Hr.			+				
60	Podozamites oblongus, Lx.	+						
61	Podozamites angustifolius ?, Hr.	+						
62	Podozamites prælongus, Lx.	+						
63	Podozamites emarginatus, Lx.	+						
64	Podozamites caudatus, Lx.	+						
65	Otozamites ? Grœnlandicus, Hr.			+				
66	Nelsonia Johnstrupi, Hr.			+				
	Coniferæ.							
67	Araucaria spathulata, Ny.	+						
68	Cuninghamites elegans, Corda.				+			
69	Cuninghamites squamosa, Hr.					+		
70	Cuninghamites oxycedrus, St.						+	
71	Cuninghamites Sternbergii, Ett.						+	
72	Cuninghamites borealis, Hr.			+				
73	Pinus Quenstedti, Hr.	+		+	+	+		
74	Pinus vaginalis, Hr.			+				
75	Pinus Staratschini, Hr.			+				
76	Pinus Upernavikensis, Hr.			+				
77	Pinus Olafiana, Hr.			+			+	
78	Abietites curvifolius, Dkr.					+		
79	Abietites Goepperti, Dkr.					+		
80	Abietites Hartigii, Dkr.					+		
81	Abietites Ernestinæ, Lx.	+				+		
82	Sequoia rigida, Hr.			+				+
83	Sequoia ambigua, Hr.			+				+
84	Sequoia Reichenbachi, Hr.	+		+	+ +	+		+
85	Sequoia pectinata, Hr.				+			

Table of Distribution of the Plants of the Cretaceous Cenomanian Formation—Continued.

Number	NAMES OF SPECIES.	Dakota Group—Kansas, Nebraska, Minnesota.	Dakota Group—Colorado, Eastern base of the Mountains.	Greenland—Schists of Atane.	Europe—Moletein, Quedlinburg.	Quader Sandstone—Harz, Bohemia.	Niederrhaoa—Saxony, Hungary.	Lower Cretaceous.
86	Sequoia fastigiata, St.	+		—	+	+		
87	Sequoia subulata, Hr			+				
88	Sequoia condita, Lx	+						
89	Sequoia? formosa, Lx	+						
90	Torreya oblanceolata, Lx		+					
91	Gelnitzia formosa, Hr				+			
92	Cyparissidium gracile, Hr			+				+
93	Glyptostrobus gracillimus, Lx			+				
94	Widdringtonites subtilis, Hr			+				
95	Frenelites Reichii, Ett			+		+	+	
96	Thuites Pfaffii, Hr							
97	Thuites crassus, Lx	+						
98	Thuites Meriani, Hr			+				+
99	Moriconia cyclotoxon, Deb			+				
100	Libocedrus cretacea, Hr			+				
101	Juniperus macileuta, Hr			+				
102	Juniperus hypnoides, Hr			+				
103	Damara borealis, Hr			+				
104	Damara microlepis, Hr			+				
105	Ginkgo primordialis, Hr			+				
106	Ginkgo multinervis, Hr			+				
107	Baiera sagittata, Hr			+				
108	Baiera leptopoda, Hr			+				
109	Baiera incurvata, Hr			+				
110	Isolepis species	+						
111	Thinfeldia Lesquereuxiana, Hr	+		+				
	Rhizanthe.							
112	Williamsonia cretacea, Hr			+				
	MONOCOTYLEDONES.							
	Glumaceæ.							
113	Arundo Groenlandica, Hr			+				
114	Phragmites cretaceus, Lx	+						
115	Culmites cretaceus, Ett						+	
	Alismaceæ.							
116	Alisma reticulata?, Hr			+				
	Coronariæ.							
117	Lamprocarplites nitidus, Hr			+				
118	Majanthemophyllum cretaceum, Hr			+				
119	Majanthemophyllum lanceolatum, Hr			+				
	Dioscoreæ.							
120	Dioscorea? cretacea, Lx	+						
	Typhaceæ.							
121	Sparganium cretaceum, Hr			+				
122	Caulinites stigmarioides, Ett						+	

¹ Heer considers these three species as synonyms.

Table of Distribution of the Plants of the Cretaceous Cenomanian Formation—Continued.

Number.	NAMES OF SPECIES.	Dakota Group—Eastman, Nebraska, Minnesota.	Dakota Group—Colorado, Eastern base of the Mountains.	Greenland—Schists of Atane.	Europe—Moletein, Quedlinburg.	Quader Sandstone—Hartz, Bohemia.	Niederschoen—Saxony, Hungary.	Lower Cretaceous.
	Scitamineæ.							
123	Zingiberites pulchellus, Hr			†				
	Pandaneæ.							
124	Pandanus Smilciæ, Stiehl				+	+		
	Palmæ.							
125	Palmacites horridus				+			
126	Flabellaria minima, Lx	+						
	DICOTYLEDONES.							
	Myricaceæ.							
127	Myrica cretacea, Hr				+			
128	Myrica Dakotensis, Lx	+						
129	Myrica obtusa, Lx	+						
130	Myriceæ semina, Lx	+						
131	Myrica Schenkiana, Hr				+			
132	Myrica Sternbergii, Lx	+		?				
133	Myrica Thulensis, Hr			+				
134	Myrica emarginata, Hr			+				
135	Myrica Zenkeri, Ett			+			+	
136	Myrica longa, Hr			+				
137	Myrica longifolia, Ung						+	
	Betulaceæ.							
138	Betula Beatriciana, Lx	+						
139	Betulites denticulata, Hr	+						
140	Phyllites betulaefolius, Lx	+						
141	Alnites grandifolius, Ny	+						
	Cupuliferæ.							
142	Dryophyllum (Quercus) latifolium, Lx	+						
143	Dryophyllum (Quercus) primordiale, Lx	+						
144	Dryophyllum (Quercus) Holmesii, Lx	+						
145	Quercus Beyrichii, Ett						+	
146	Quercus Dakotensis, Lx	+						
147	Quercus hexagona, Lx	+						
148	Quercus Ellsworthiana, Lx	+						
149	Quercus Westfalica, Hos. & v. d. M			+				
150	Quercus Rinklaae, Hr			+				
151	Quercus Warningiana, Hr			+				
152	Quercus ferox, Hr			+				
153	Quercus hierncifolia, Hos. & v. d. M			+				
154	Quercus thulensis, Hr			+				
155	Quercus troglodites, Hr			+				
156	Quercus poranoides, Lx	+						
157	Quercus Morrisoniana, Lx		+					
158	Quercus salicifolia, Ny	+						
159	Quercus antiqua, Ny	+						
160	Quercus sinuata, Ny	+ ?						
161	Castanea Hausmanni, Dkr					+		

Table of Distribution of the Plants of the Cretaceous Cenomanian Formation—Continued.

Number.	NAMES OF SPECIES.	Dakota Group—Kansas, Nebraska, Minnesota.	Dakota Group—Colorado, Eastern base of the Mountains.	Greenland—Schists of Atane.	Europe—Moletein, Quedlinburg.	Quader Sandstone—Harz, Bohemia.	Niederschöna—Saxony, Hungary.	Lower Cydaceous.
162	Fagus prisca, Ett						+	
163	Fagus polyclada, Lx	+						
164	Fagus cretacea, Ny	+						
	Salicineæ							
165	Salix nervillosa, Hr	+						
166	Salix Goetziana, Hr				+			
167	Salix protecefolia, Lx	+						
168	Salix Meckii, Ny	+						
169	Salix flexnosa, Ny	+						
170	Salix cuneata, Ny	+						
171	Salix Hartigii, Dkr					+		
172	Populus litigiosa, Hr	+						
173	Populus cyclophylla, Hr	+						
174	Populus elliptica, Ny	+						
175	Populus microphylla, Ny	+						
176	Populus ? cordifolia, Ny	+						
177	Populus Berggreni, Hr			+				
178	Populus amissa, Hr			+				
179	Populus hyperborea, Hr			+				
180	Populus stygia, Hr			+				
181	Populus primæva, Hr							+
182	Populites Lancastriensis, Lx	+						
183	Populites elegans, Lx	+						
	Plataneæ.							
184	Platanus Newberryana, Hr	+						
185	Platanus obtusiloba, Lx	+						
186	Platanus primæva, Lx	+						
187	Platanus Heerii, Lx	+		+				
188	Platanus diminutiva, Lx	+						
	Styraciflux.							
189	Liquidambar integrifolium, Lx	+						
	Moreæ.							
190	Ficus protogæa, Ett						+	
191	Ficus protogæa, Hr						+	
192	Ficus Geinitzii, Ett						+	
193	Ficus bumelioides, Ett						+	
194	Ficus primordialis, Hr	+						
195	Ficus Mohliana, Hr				+			
196	Ficus Krausiana, Hr				+			
197	Ficus Halliana, Lx	+						
198	Ficus Beckwithii, Lx		+					
199	Ficus angustata, Lx	+						
200	Ficus Magnoliæfolia, Lx		+					
201	Ficus Glascœna, Lx	+						
202	Ficus distorta, Lx	+						
203	Ficus laurophylla, Lx	+						
204	Ficus Atanepa, Hr			+				

C F 7

Table of Distribution of the Plants of the Cretaceous Cenomanian Formation—Continued.

Number.	NAMES OF SPECIES.	Dakota Group—Kansas, Nebraska, Minnesota.	Dakota Group—Colorado, Eastern base of the Mountains.	Greenland—Schists of Atane.	Europe—Moletein Quedlinburg.	Quader Sandstone—Hartz, Bohemia.	Niederschoena—Saxony, Hungary.	Lower Cretaceous.
205	Ficus crassipes, Hr.			+				
206	Ficus Hollandiana, Hr.			+				
	Artocarpeæ.							
207	Artocarpidium cretaceum, Ett.						+	
	Urticaceæ.							
208	MacClintockia cretacea, Hr.			+				
209	MacClintockia appendiculata, Hr.			+				
	Daphneæ.							
210	Daphnites Goepperti, Ett.						+	
	Proteaceæ.							
211	Protea Haidingeri, Ett.						+	
212	Proteoides lancifolius, Hr.	+			+			
213	Proteoides ilicoides, Hr.				+			
214	Proteoides Grevilliæformis, Hr.	+	+					
215	Proteoides Daphnogenoides, Hr.	+						
216	Proteoides acuta, Hr.	+						
217	Proteoides longus, Hr.			+				
218	Proteoides crassipes, Hr.			+				
219	Proteoides vexans, Hr.			+				
220	Conospermites hakeæfolius, Ett.						+	
221	Rapala primæva, Ett.						+	
222	Banksia prototypus, Ett.						+	
223	Banksia longifolia, Ett.						+	
224	Lomatia saportanea, Lx.	+	+					
225	Lomatites palæo-ilex, Ett.						+	
226	Dryandroides Zenkeri, Ett.						+	
227	Dryandroides latifolius, Ett.						+	
	Laurineæ.							
228	Laurus Nebrascensis, Lx.	+						
229	Laurus macrocarpa, Lx.	+						
230	Laurus proteæfolia, Lx.	+	+					
231	Laurus modesta, Lx.			+				
232	Laurus cretacea, Ett.						+	
233	Laurus plutonia, Hr.			+				
234	Laurus angusta, Hr.			+				
235	Laurus Hollæ, Hr.			+				
236	Laurus Odini, Hr.			+				
237	Persea Leconteana, Lx.	+						
238	Persea Sternbergii, Lx.	+						
239	Daphnogene primigenia, Ett.						+	
240	Cinnamomum Scheuchzeri, Hr.	+						
241	Cinnamomum Heerii, Lx.	+						
242	Cinnamomum Sezannense, Wat.			+				
243	Oreodaphne cretacea, Hr.	+						
244	Sassafras arctica, Hr.			+				
245	Sassafras Mudgei, Lx.	+						
246	Sassafras acutilobum, Lx.	+						

Table of Distribution of the Plants of the Cretaceous Cenomanian Formation—Continued.

Number.	NAMES OF SPECIES.	Dakota Group—Kansas, Nebraska, Minnesota.	Dakota Group—Colorado, Eastern base of the Mountains	Greenland—Schists of Atane.	Europe—Moletein, Quedlinburg.	Quader Sandstone—Hartz, Bohemia.	Niederschœna—Saxony, Hungary.	Lower Cretaceous
247	Sassafras (Araliopsis) cretaceum, Ny	+						+
248	Sassafras (Araliopsis) obtusum, Lx	+						
249	Sassafras (Araliopsis) mirabile, Lx	+						
250	Sassafras (Araliopsis) dissectum, Lx	+						
251	Sassafras (Araliopsis) recurvatum, Lx	+		+				
252	Sassafras (Araliopsis) platanoides, Lx	+						
253	Sassafras (Araliopsis) deformatum, Lx	+						
254	Daphnophyllum Fraasii, Hr				+			
255	Daphnophyllum crassinervium, Hr				+			
256	Daphnophyllum ellipticum, Hr				+			
	Apocyneæ.							
257	Apocynophyllum cretaceum, Ett						+	
	Asarineæ.							
258	Aristolochia dentata, Hr	+						
	Myrsineæ.							
259	Myrsine borealis, Hr			+				
	Diospyrineæ							
260	Sapotacites Haydenii, Ny	+						
261	Diospyros primæva, Hr	+		+				
262	Diospyros ambigua, Lx	+						
263	Diospyros rotundifolia, Lx	+						
264	Diospyros prodromus, Hr			+				
	Ericaceæ.							
265	Andromeda Parlatorii, Hr	+		+				
266	Andromeda Pfaffiana, Hr			+				
267	Dermatophyllites borealis, Hr			+				
268	Dermatophyllites acutus, Hr			+				
	Asclepiadeæ.							
269	Acerates arctica, Hr			+				
	Araliaceæ.							
270	Aralia formosa, Hr	+			+	+		
271	Aralia Saportanea, Lx	+						
272	Aralia Towneri, Lx	+	+					
273	Aralia Ravniana, Hr			+				
274	Aralia quinquepartita, Lx	+						
275	Aralia emarginata, Lx	+						
276	Aralia concreta, Lx	+						
277	Aralia tenuinervis, Lx	+						
278	Aralia radiata, Lx	+						
279	Aralia Groenlandica, Hr			+				
280	Panax cretaceum, Hr			+				
281	Hedera ovalis, Lx	+						
282	Hedera primordialis, Hr			+				
283	Hedera Schimperi, Lx	+						
284	Hedera cuneata, Hr			+				
285	Hedera platanoides, Lx	+						

Table of Distribution of the Plants of the Cretaceous Cenomanian Formation—Continued.

Number	NAMES OF SPECIES.	Dakota Group—Kansas, Nebraska, Minnesota.	Dakota Group—Colorado, Eastern base of the Mountains.	Greenland—Schluss of Atane.	Europe—Moletein, Quedlinburg.	Quader Sandstone—Hartz, Bohemia.	Niederschoena Saxony, Hungary.	Lower Cretaceous.
	Ampelideæ.							
286	Cissites insignis, Hr	+						
287	Cissites formosus, Hr			+				
288	Cissites Harkerianus, Lx	+						
289	Cissites affinis, Lx	+		+				
290	Cissites acuminatus, Lx	+						
291	Cissites Heerii, Lx	+						
292	Cissites salisburiæfolius, Lx	+						
293	Chondrophyllum orbiculatum, Hr			+				
294	Chondrophyllum Nordenskiöldi, Hr			+				
295	Chondrophyllum hederæformis, Hr						+	
296	Ampelophyllum attenuatum, Lx	+						
297	Ampelophyllum ovatum, Lx	+						
	Hamamelideæ.							
298	Hamamelites Kansaseanus, Lx	+						
299	Hamamelites tenuinervis, Lx	+						
300	Hamamelites quadrangularis, Lx	+						
301	Hamamelites ? cordatus, Lx	+						
	Corneæ.							
302	Cornus Forchammeri, Hr			+				
	Magnoliaceæ.							
303	Magnolia alternans, Hr	+	+	+				
304	Magnolia Isbergiana, Hr			+				
305	Magnolia Capellini, Hr	+	+	+				
306	Magnolia obtusata, Hr			+				
307	Magnolia speciosa, Hr		+		+	+		
308	Magnolia amplifolia, Hr				+	+		
309	Magnolia tenuifolia, Lx	+						
310	Magnolia obovata, Ny	+						
311	Magnolia species, fruit		+					
312	Liriodendron Meekii, Hr	+		+				
313	Liriodendron primævum, Ny	+						
314	Liriodendron intermedium, Lx	+						
315	Liriodendron giganteum, Lx	+						
316	Liriodendron acuminatum, Lx	+						
317	Liriodendron cruciforme, Lx	+						
318	Liriodendron semi-alatum, Lx	+						
319	Liriodendron pinnatifidum, Lx	+						
320	Liriophyllum Beckwithii, Lx		+					
321	Liriophyllum populoides, Lx	+	+					
322	Liriophyllum cordatum, Lx		+					
323	Carpites liriophylli ?, Lx		+					
	Anonaceæ.							
324	Anona cretacea, Lx	+						
	Ranunculaceæ.							
325	Dewalquea insignis, Hos. & V. d. M			+				
326	Dewalquea Groenlandica, Hr			+				

Table of Distribution of the Plants of the Cretaceous Cenomanian Formation—Continued.

Number.	NAMES OF SPECIES.	Dakota Group—Kansas, Nebraska, Minnesota.	Dakota Group—Colorado, Eastern base of the Mountains.	Greenland—Schists of Atane.	Europe—Moletein, Quedlinburg.	Quader Sandstone—Hartz, Bohemia.	Niedershœna—Saxony, Hungary.	Lower Cretaceous.
	Menispermaceæ.							
327	Menispermites obtusifolius, Lx	+						
328	Menispermites Salinensis, Lx	+						
329	Menispermites acerifolius, Lx	+						
330	Menispermites populifolius, Lx	+						
331	Menispermites cyclophyllus, Lx	+						
332	Menispermites grandis, Lx	+						
333	Menispermites acutilobus, Lx	+						
334	Menispermites dentatus, Hr			+				
335	Menispermites ovalis, Lx	+						
336	Menispermites borealis, Hr			+				
	Nymphæaceæ.							
337	Nelumbium arcticum, Hr			+				
	Malvaceæ.							
338	Sterculia obtusiloba, Lx	+						
339	Sterculia aperta, Lx	+						
340	Sterculia lugubris, Lx	+	+					
	Tiliaceæ.							
341	Grewiopsis Haydenii, Lx	+						
342	Apeibopsis Thomsoniana, Hr			+				
	Aceraceæ.							
343	Sapindus prodromus, Hr			+				
344	Sapindus Morrisoni, Lx		+	+				
345	Acer antiquum, Ett	+					+	
346	Acerites pristinus, Ny	+						
347	Negundoides acutifolius, Lx	+						
	Frangulineæ.							
348	Celastrophyllum ensifolium, Lx	+						
349	Celastrophyllum lanceolatum, Ett						+	
350	Celastrophyllum integrifolium, Ett						+	
351	Celastrophyllum obtusum, Hr			+				
352	Paliurus membranaceus, Lx	+						
353	Ilex strangulata, Lx	+						
354	Ilex antiqua, Hr			+				
355	Rhamnus Œrstedi, Hr			+				
356	Rhamnus prunifolius, Lx	+						
357	Rhamnus tenax, Lx	+						
358	Rhamnus acuta, Hr			+				
	Myrtaceæ.							
359	Eucalyptus Geinitzi, Hr			+	+			
360	Eucalyptus borealis, Hr			+				
361	Myrtophyllum parvulum, Hr			+				
362	Myrtophyllum pusillum, Hr					+		
363	Myrtophyllum Schübleri, Hr			+				
364	Metrosideros peregrinus, Hr			+				
365	Callistemophyllum Heerii, Ett						+	

Table of Distribution of the Plants of the Cretaceous Cenomanian Formation—Continued.

Number.	NAMES OF SPECIES.	Dakota Group—Kansas, Nebraska, Minnesota.	Dakota Group—Colorado, Eastern base of the Mountains.	Greenland—Schiste of Atane.	Europe—Moletein, Quedlinburg.	Quader Sandstone—Hartz, Bohemia.	Niederschœna—Saxony, Hungary.	Lower Cretaceous.
	Columniferæ.							
366	Pterospermites cordifolius, Hr			+				
367	Pterospermites auriculatus, Hr			+				
	Juglandeæ.							
368	Juglans? Debeyana, Hr	+						
369	Juglans? crassipes, Hr						+	
370	Juglans? arctica, Hr			+				
	Anacardiaceæ.							
371	Rhus cretacea, Hr				+			
372	Rhus microphylla, Hr			+				
373	Phyllites rhoifolius, Lx		+					
374	Anacardites amissus, Hr			+				
	Pomaceæ.							
375	Pyrus? cretacea, Ny	+						
	Amygdaleæ.							
376	Prunus? cretacea, Lx	+						
	Leguminosæ.							
377	Colutea primordialis, Hr			+				
378	Colutea longæana, Hr			+				
379	Colutea valde-inæqualis, Hr			+				
380	Dalbergia kinkiana, Hr			+				
381	Dalbergia hyperborea, Hr			+				
382	Palæocassia angustifolia, Ett			+			+	
383	Palæocassia lanceolata, Ett			+			+	
384	Inga Cottai, Ett						+	
385	Cassia Ettingshauseni, Hr			+				
386	Cassia antiquorum, Hr			+				
387	Leguminosites prodromus, Hr			+				
388	Leguminosites ovalifolius, Hr			+				
389	Leguminosites insularis, Hr			+				
390	Leguminosites atanensis, Hr			+				
391	Leguminosites corollinoides, Hr			+				
392	Leguminosites amissus, Hr			+				
393	Leguminosites (legumen), Lx		+					
394	Leguminosites macilentus, Hr			+				
395	Leguminosites orbiculatus, Hr			+				
396	Leguminosites Dalageri, Hr			+				
	Genera and species of uncertain relation.							
397	Credneria macrophylla, Hr				+			
398	Credneria integerrima, Zenk			+		+	+	
399	Credneria denticulata, Zenk						+	
400	Credneria subtriloba, Zenk						+	
401	Credneria acuminata, Hmp						+	
402	Credneria subserrata, Hmp						+	
403	Credneria tricuminata, Hmp						+	
404	Credneria Schneideriana, Gœpp						+	

Table of Distribution of the Plants of the Cretaceous Cenomanian Formation—Continued.

Number.	NAMES OF SPECIES.	Dakota Group—Kansas, Nebraska, Minnesota.	Dakota Group—Colorado, Eastern base of the Mountains.	Greenland—Schists of Atane.	Europe—Moletein, Quedlinburg.	Quader Sandstone—Hartz, Bohemia.	Niederschoena—Saxony, Hungary.	Lower Cretaceous.
405	Creduerin Sternbergii, Brgt.				+			
406	Creduerin cuneifolia, Bronn						+	
407	Creduerin Geinitziana, Ung						+	
408	Creduerin grandidentata, Ung				+		+	
409	Creduerin species, Hr			+				
410	Aspidiophyllum trilobatum, Lx	+						
411	Aspidiophyllum platanifolium, Lx	+						
412	Aspidiophyllum dentatum, Lx	+						
413	Protophyllum Sternbergii, Lx	+						
414	Protophyllum Leacooteanum, Lx	+						
415	Protophyllum Nebrascense, Lx	+						
416	Protophyllum quadratum, Lx	+						
417	Protophyllum minus, Lx	+						
418	Protophyllum multinerve, Lx	+						
419	Protophyllum rugosum, Lx	+						
420	Protophyllum Haydenii, Lx	+						
421	Protophyllum crednerioides, Lx	+						
422	Protophyllum Mudgei, Lx	+						
423	Anisophyllum semi-alatum, Lx	+						
424	Ereunophyllum fimbriatum, Lx	+						
425	Phyllites Vanonae, Hr	+						
426	Phyllites obcordatus, Ny	+						
427	Phyllites rhomboideus, Lx	+						
428	Phyllites cutinus, Lx	+						
429	Phyllites umbonatus, Lx	+						
430	Phyllites amorphus, Lx	+						
431	Phyllites linguaeformis, Hr			+				
432	Phyllites laevigatus, Hr			+				
433	Phyllites longepetiolatus, Hr			+				
434	Phyllites granulatus, Hr			+				
435	Phyllites incurvatus, Hr			+				
436	Phyllites celastroides, Hr				+			
437	Phyllites ramosinervis, Hr				+			
438	Tetraphyllum oblongum, Hr			+				
439	Carpolithes? species, Lx	+						
440	Carpolithes? scrobiculatus, Hr			+				
441	Carpolithes? cretaceus, Ett						+	
442	Caudex spinosus, Lx	+						

THE RELATIONSHIP OF THE FLORA OF THE DAKOTA GROUP.

-

In comparing first the Flora of the Dakota Group to plants described by Heer from Kome, referable to the lowest Cretaceous or Neocomian formation, the table of distribution indicates an extremely great difference in the characters of the constituents. Two species only are common to both these groups of plants: *Gleichenia Nordenskiöldi*, a fern, and *Sequoia Reichenbachi*, a Conifer. These species are of predominant and persistent Jurassic types, remnants of old epochs. The single dicotyledonous species discovered in the group of plants of Kome, *Populus primæva*, belongs to the section of the coriaceous poplars, represented at Atane by two other species. No poplar of this section has been observed as yet among the vegetable remains of the Dakota Group. This last flora is, therefore, without affinity to that of Kome. But with the flora of Atane that of the Dakota Group has a marked degree of affinity, 15 species of plants being common to both. They are: *Pinus Quenstedti, Sequoia Reichenbachi, S. fastigiata, Thinfeldia Lesquereuxiana, Platanus Heerii, Ficus Mohliana, Sassafras recurvatum, Diospyros primæva, Andromeda Parlatorii, Cissites affinis, Magnolia alternans, Magnolia Capellini, Liriodendron Meekii, Sapindus Morrisoni.* Besides these, *Thuites crassus* and *Myrica Sternbergii* of the Dakota Group are so closely allied to *T. Pfaffii* and *M. Thulensis* of Atane that these forms, described under different specific names, appear to be mere varieties; and the same can be said of *Ficus protogæa* and *Aralia Rœniana* of Atane, which, as far as can be surmised in comparing figures and descriptions, appear identical with *Ficus Beckwithii* and *Aralia Towneri* of the Dakota Group. The relationship is the more remarkable as the affinities are not limited to one or a few peculiar sections of the vegetable kingdom, but refer to plants of most of the divisions known in the flora of the present epoch, at least in that of the temperate regions. Of the 65 genera to

which the plants of the Dakota Group have been referred, 40 are repre-
sented at Atane; and in them (besides Ferns, Conifers, Monocotyledons)
there are, in the Dicotyledons, *Magnoliaceæ, Anonaceæ, Menispermaceæ,
Vitaceæ, Sapindaceæ, Araliaceæ,* under the subdivision of the Polypetalous;
Leguminosæ, Ericaceæ, Ebenaceæ, in the Monopetalous; *Hamamelaceæ,
Cornaceæ, Rhamnaceæ, Urticaceæ (Moreæ Juglandeæ,* etc. *),* in the Apetalous.
Hence the relation of these floras is, so to speak, general.   There is only a
marked difference in the number of species represented in a few groups.
Atane, for example, has 35 species of ferns and 28 of Conifers, while only
6 ferns and 9 Conifers are known from the Dakota Group.   This last flora
has a large number of species in the genera *Salix, Platanus, Sassafras,
Aralia, Liriodendron, Menispermites, Protophyllum,* while Atane has predominance of species in *Magnolia,* in the *Myrtaceæ, Pterospermites, Rhus,*
and especially in the *Leguminosæ,* of which 18 species are described by
Heer, while only one is known from the Dakota Group.   But these differences merely show the influence of local circumstances, lower temperature,
more open ground perhaps for the plants of Atane, where ferns and
Leguminosæ are more abundantly distributed than in forests of large-leafed
trees, like those of which the flora of the Dakota Group is especially
composed.

As Kome and Atane have in common 8 species of Ferns and Gymnos-
perms, of which two only have been found in the Dakota Group, it might
be supposed that the Atane flora is older than that of the Dakota Group.
The characters of the Dicotyledonous plants lead to a different conclusion: for some of these plants of Atane are identical or very closely related
to species of the upper Cretaceous, or Senonian, while none of them have
been observed in the Dakota Group; *Quercus Westfalica* and *Q. hieracifolia,*
recorded by Heer in the flora of Atane, are described from the Senonian of
Europe; two species of *Dewalquea,* also recognized by Heer in the plants
of Atane, are found in the upper Cretaceous of Belgium and the Paleocene
of France, while *Cinnamomum Sezannense,* which Heer has also found in
the plants of Atane, is lower Eocene in France.   Therefore, it is evident
that the formation of Atane is somewhat more recent than that of the
Dakota Group, apparently an upper stage of the same.

The degree of relationship of the Dakota Group flora with that of the

Cenomanian of Europe in divers localities indicated in the table, is the least distinctly marked with Quedlinburg. From this place Heer has described 20 species, 3 of which only—*Gleichenia Kurriana, Sequoia Reichenbachi,* and *Proteoides lancifolius*—are identified in the Dakota Group. The stage of the Quedlinburg beds is not positively determined. While some geologists refer it to the Cenomanian, Goeppert considers it as lower Senonian, or as a formation more recent than that of the Cretaceous of Kansas. It has a *Credneria* (*C. integerrima,* Zenk.), also found at Atane. The flora of Moletein offers, in nearly the same number of species (18), more definite points of affinity with that of the Dakota Group in 7 identical species, 3 of which are dicotyledonous: *Ficus Mohliana, Aralia formosa,* and *Magnolia speciosa.* The Moletein formation is generally admitted as equivalent to that of the lower Quader sandstone of Germany, from which at different localities in the Hartz and in Bohemia 30 species of plants have been described. Of these, also, 8 are found in the Dakota Group. Hence the marked analogy in the components of these floras authorizes the conclusion of equivalency of the age of the Dakota Group with that of the Quader sandstone of Germany, which is as positively determined as Cenomanian by its animal fossils as the Dakota Group is recognized as Middle Cretaceous by the invertebrate remains which abound in the strata of the Fort Benton Group, immediately overlying it.

We may have an opportunity to see in the characters of the plants further described in this volume, from the different stages of the Tertiary, some of the types of the Dakota Group reappearing through subsequent periods, especially in the Miocene. But this cannot in any way nullify the originality of these types, and what is said above sufficiently proves that if the Dakota Group has in its flora some plants closely allied to Miocene species, and also to plants living at the present time, the Cretaceous age of the group is positively fixed.

FLORA OF THE LARAMIE GROUP.

The age of the Laramie Group of Hayden is not yet definitively determined. The remains of fossil plants, abundantly procured from this formation, especially at Golden, Black Buttes, and Point of Rocks, have been recognized by botanists as pertaining to a flora mostly composed of Tertiary types, while, according to zoölogists, the fauna of the same formation is Cretaceous in character. Though the question has already been discussed at length and considered under diverse points of view, my own opinion being given in the preceding volume of the "U. S. Geol. Rep.," vii. pp. 338–352, in F. V. Hayden's "Ann. Rep.," 1872 to '74, etc., it is proper briefly to present here some new facts bearing on the subject, and to note the conclusions which may be derived from them.

1st. The flora of the Laramie Group has a relation, remarkably well defined, with that of Sézanne. This relation becomes still more distinctly shown by the few species of plants which have recently been added to it and are described below. The flora is not vague or indefinite in its character; its types are clear and precise; those which are limited to the formation are found in the divers localities where the remains of plants have been discovered, the relation of some others is with plants of a higher stage, especially with those of the Miocene; very few are Cretaceous, and these are mostly represented by persistent species which, derived from the Jurassic, have passed through the intervening period to the present epoch.

Though the geological surveys of the Government have not sent me from the Laramie Group any specimens of fossil plants to be examined and described in this volume, I have had the opportunity of looking over a large collection of plant remains obtained at Golden for the Museum of Comparative Zoölogy of Cambridge. They mostly represent species already known. Of the new ones, none are referable to Cretaceous types; they are still more generally allied to those of Sézanne. This does not imply that

the flora of the Laramie is positively identical in its geological horizon
with that of Sézanne. There are marked differences in the general characters of the vegetable groups. The flora of the Laramie, for example,
has a remarkable predominance of species of palms, while these are, on
the contrary, very rare at Sézanne. As the palms have their origin, as
far as known, in the middle Cretaceous, where they have been observed in
very rare remains, limited to one or two species, and as their development
has been gradually progressing through the more recent formations, this
fact, or the abundance of remains of palms in the flora of the Laramie, gives
to it a somewhat more recent aspect than that of Sézanne, where the absence
of palms, however, may have resulted from mere local circumstances.

2d. Some time ago the members of one of the scientific expeditions of
Princeton College discovered and collected in Wyoming a number of fine
specimens of fossil plants referable, by their characters, to a stage of the
Cretaceous more recent than the Cenomanian Dakota Group. As far as
can be judged by a preliminary examination, the species, mostly *Quercites*
and *Araliaceæ*, are related by identical types, even by some identical species,
to the flora of the Senonian, as it is known in Germany by the plants
published by Hosius and Von der Mark, and in Belgium by those of Debey.
They have also a degree of affinity, though less distinct, with those of the
Marnes Heersiennes of Gelinden, a formation which, in France, constitutes
part of the series of the Sables de Bracheux or of the London clay, etc.,
the lowest part of the Tertiary system, or Eocene, as it is generally admitted
to be by European geologists. The plants of Gelinden, partly Senonian in
their characters, are related to the Sézanne flora by one identical species
and a number of others of generic or typical affinity. Hence we see now,
in the floras of the North American Continent, from the Cenomanian to the
Eocene of the Laramie, a succession of vegetable groups corresponding to
the European series, with the exception only of the flora of Gelinden in
the Sables of Bracheux, not yet discovered on this continent. According
to French geologists the Sézanne beds are comprised in the Pisolitic limestone, a formation superior to the Sables of Bracheux, and hence more
distinctly referable to the Tertiary.

3d. A memoir published by Professor Cope on the horizon of extinct vertebrates of Europe and North America[1] contains very valuable and interesting documents, which really show that the evidence afforded as to the age of the Laramie Group both by the remains of animals and by those of plants is not far discordant. In the table indicating the correlation of all the formations from the lowest to the more recent (pp. 50 and 51 of the memoir quoted above) the horizon of the Sézanne flora, or the Pisolitic limestone, is not separately indicated, but is probably in what the author calls the Puerco stage, hypothetically identified with the Thanetian, or lower Eocene; the whole Puerco and Laramie on one side, and the Sables of Bracheux on the other, being marked as Post-Cretaceous. Now the relation and difference between the vertebrates of the Laramie and those of the Sables of Bracheux is established by Professor Cope as follows: "The genera of *Dinosauria (Palæoscincus, Cionodon, Diclonius, Monoclonius, Dysganus)*, which constitute a predominant type in the Laramie Group, have not been found in any other part of the world. Mingled with them were species of crocodiles and turtles of indifferent characters, while a number of other forms existed which had a limited range in time, and hence are important indications of stratigraphic position. Such are the genera *Myledaphus* (Cope) and *Clastes* (Cope), which have been found also near Rheims, France, by Dr. Lemoine, in the Sables de Bracheux, which are regarded as the lowest Tertiary. Such is the curious Saurian type *Champsosaurus* (Cope), *Simœdosaurus* (Grev.), and the turtle genus *Compsemys* (Leidy), which Lemoine finds a little higher up in the series in the conglomerate of Cerny, which is the lower part of the Suessonian. In France, a genus of the Laramie, *Polythorax*, extends into the Lignite or upper *Coryphodon* beds of the Suessonian. Thus the Laramie is intercalated in its characters between the Cretaceous period on one hand and the Tertiary on the other, and its fauna includes genera and orders of both great series."

Admitting the exposition of the characters of the strata as made by the celebrated author of the notice, it may be observed that, from the table which follows the above remark, all the genera common to the Sables of Bracheux and the Laramie Group forcibly indicate relationship to the

---

[1] The relation of the horizon of extinct vertebrata of Europe and North America. "U. S. Geol. & Geog. Survey" (Hayden), Bull. v, No. 1.

Tertiary, even to strata above the Eocene. The other genera, as remarked by Professor Cope, are *Dinosaurian* of Mezozoic types, but are without any representatives in Europe; hence they can only be used as hypothetically implying reference of the Laramie Group to the Post-Cretaceous. For they have never been found anywhere but in America, while the reference of the Laramie to the Tertiary age is based on the positive evidence of species or genera represented in that formation both in Europe and America.

Professor Heer, in the VIth volume of the "Arctic Flora," has examined the question from the same point of view. After remarking that the Tertiary character of the fossil plants of the Laramie Group, confirmed by that of the mollusks, had rightly forced me to recognize it as Tertiary, he adds that the discovery at Black Buttes of *Agathaumas sylvestris*, a Dinosaurian, had been considered by zoölogists as sufficient authority for the admission not only of Black Buttes but of the whole Laramie Group into the Cretaceous; this from the dogma that Dinosaurians have disappeared with the Cretaceous. That a Saurian, he says, has been found only at that locality, is no reason for recognizing it as a Cretaceous species, but the only conclusion which can be drawn from the fact is, that until now it has been supposed that the Dinosaurian type had died in the Cretaceous, while animals of this kind have permitted some of their offspring to live still in the Tertiary. And, indeed, in regard to that, other groups of Saurians, like the crocodile, have lived in far different periods. Therefore the *Agathaumas* of Black Buttes is not proof at all that at that locality a Tertiary flora was existing at the same time as a Cretaceous fauna, as admitted by Professor Cope; for a single animal does not constitute a fauna any more than a fragment of plant could constitute a flora. Added to this, it is also well to remark that at Black Buttes, in a stratum immediately above the bed where the remains of *Agathaumas* were found, a fish, *Celastes*, four species of turtles, an alligator, and a mammal have been discovered, and that all these animals are undoubtedly Tertiary.[1]

4th. The Laramie formation is a land or fresh-water formation. If sufficient proof of this fact was not given by the remains of plants and the numerous coal deposits found at divers stages over its whole extent.

[1] O. Heer, Beitrage zur Miocene Flora von North Canada, p. 7, in Flora fossilis Arctica, vol. vi. part 2.

the molloscan fauna would offer an incontestable evidence. Professor C. A. White, in a paper lately published,[1] writes as follows: "The invertebrate fauna of the Laramie Group is wholly different from that of any of the Marine Cretaceous formations, with one of which some writers have confounded it. It contains no true marine type of any kind, but it does contain many brackish-water molluscan forms, and also the remains of many fresh and land mollusks. The fauna characterizes a great widespread geological group of strata in the most distinct and unequivocal manner, several of its molluscan species now being known to occur at localities more than a thousand miles apart." After remarking on the erroneous statements in the text-book of Geology by Professor Geikie, and on the assertion of Professor J. P. Stevenson upon the presence of marine strata of the Fox Hills Group alternating with those of the Laramie. Professor White adds: "That any true Laramie strata ever alternate with those of the Fox Hills Group, or any other Marine Cretaceous Group, or that any true marine fossils were ever collected from any strata of the Laramie Group, I cannot admit. I regard all such statements as the result of a misunderstanding of the stratigraphical geology of a region in which such observations are said to have been made."

These remarks agree entirely with those I have had opportunity to make in my researches on the flora of the Laramie Group.[2] The flora, like the invertebrate fauna, is, on the whole, of a peculiar character, uniformly distributed over the whole extent of the formation, and free from any types or characters relating it to the Cretaceous flora. As the Laramie Group has never been subjected to submersion in the deep sea, the few remains of Dinosaurians found in it are derived from low marine lagoons penetrating into the land, and cannot impress the formation with the Cretaceous character. This being the case, it is not at all surprising to find remains of marine animals of Cretaceous types with remains of plants of Tertiary age, not more than to find the bones of the marine saurian *Agathaumas* of Black Buttes enveloped in a mass of dicotyledonous leaves, some of them even glued to the bones, and petrified with them

[1] Late observations concerning the Molluscan Fauna and the Geographical extent of the Laramie Group, "Amer. Journ. of Sci.," 3d series, vol. xxv, p. 206 (1883).
[2] "American Journal of Science," 3d Ser., 1874, vol. xxv, pp. 546-557.
C F 8

in such a way that they cannot be separated without breaking the specimens. This fact positively indicates the cause of the distribution of some
remains of Cretaceous animals as merely casual, without relation to the
nature and the progressing development of the formation.

As has already been remarked, the external aspect of the species of
different groups treated in vol. vii is an obstacle to the easy comprehension
of the character of each group. It is, therefore, advisable to have now,
separately, all the species of the Eocene flora exposed in a table, with
their relation indicated. This will render more clear the deductions
which, as said above, have been derived from the character of the flora in
the "U. S. Geol. Rep.," vol. vii.[1]

---

[1] This quotation refers to vol. vii of the " U. S. Geological Survey of the Territories," by Dr. F. V. Hayden
(1878).

Table of Distribution of the Species of the Laramie Group—Continued.

NAMES OF SPECIES.	Raton Mountains, Placize, Colorado, Henry's Fork, Barrell's Springs, Fort Ellis, Spring Cañon, Black Buttes, Point of Rocks, Yellowstone Lake.	Mississippi Eocene.	Green River Group—Oligocene.	Carbon, Alaska, etc.—Miocene.	Sézanne—Eocene.	Bornstädt, Mt. Promlus—Oligocene.	Miocene.
GYMNOSPERMÆ.							
CYCADEÆ.							
Zamiostrobus? mirabilis, Lx	Col						
CONIFERÆ.							
Thuites? complanatus, Lx	P. of R					Bel	
Sequoia Langsdorfii, Brgt	B. B., Col		Id				Id
Sequoia brevifolia, Hr	P. of R., B. Sp						Id
Sequoia longifolia, Lx	Col., P. of R						
Sequoia acuminata, Lx	B. D.						
Sequoia biformis, Lx	P. of R						
Abietites dubius, Lx	B., F. E., etc						
Abietites setiger, Lx	Sp. C						
Salisburia polymorpha, Lx	F. E.	Id					
MONOCOTYLEDONES.							
GLUMACEÆ.							
Arundo? obtusa, Lx	Col					Bel	
Phragmites Œningensis, Al. Br	Col						Id
Phragmites Alaskana, Hr	Sp. C			Id			Id
Carex Berthoudi, Lx	Col						
SMILACINEÆ.							
Smilax grandifolia, Ung	Col			Id			
SCITAMINEÆ.							
Zingiberites dubius, Lx	Col						
HYDROCHARIDEÆ.							
Ottelia Americana, Lx	P. of R					Bel	
NAJADEÆ.							
Caulinites sparganioides, Lx	Col., B. B., Sp. C						
Caulinites fecundus, Lx	Col						
LEMNACEÆ.							
Lemna scutata, Daws	P. of R						
ARACEÆ.							
Pistia corrugata, Lx	P. of R						
INCERTÆ SEDIS.							
Eriocaulon? porosum, Lx	Col						
Phyllites improbatus, Lx	B. B.						
PALMÆ.							
Flabellaria Zinckeni, Hr	Col., B. S.					Id	
Flabellaria coconica, Lx	Col					Bel	
Sabalites Grayanus, Lx	Col., P. of R	Id					
Sabalites Campbellii, Ny	R. M., Col					Bel	
Sabalites fructifer, Lx	Col						

Table of Distribution of the Species of the Laramie Group—Continued.

NAMES OF SPECIES.	AMERICAN.				EUROPEAN.		
	Raton Mountains, Placfire, Colorado, Henry's Fork, Barrell's Springs, Fort Ellis, Spring Cañon, Black Buttes, Point of Rocks, Yellowstone Lake.	Mississippi Eocene.	Green River Group—Oligocene.	Carbon, Alaska, etc.—Miocene.	Sézanne—Eocene.	Bornstadt, Mt. Prom-ina—Oligocene.	Miocene.
Geonomites Goldianus, Lv	Col					Rel	
Geonomites Schimperi, Lx	Y S. Lake					Rel	
Geonomites tenuirachis, Lx	R. M						
Geonomites Ungeri, Lx	R. M					Rel	
Oredoxites plicatus, Lx	Col						
Palmocarpus compositus, Lx	P						
Palmocarpus Mexicanus, Lx	P						
Palmocarpus communis, Lx	R., Col , B. B						
Palmocarpus truncatus, Lx	Col						
Palmocarpus corrugatus, Lx.	Col						
Palmocarpus subcylindricus, Lx	Col						
DICOTYLEDONES.							
AMENTACEÆ.							
Myrica Torreyi, Lx	B. B.					Rel	
Myrica? Lessigii, Lx	Col		Rel				Rel
Myrica ? pungens, Lx	Col				Rel		
Betula gracilis, Ludw	Col						Id
CUPULIFERÆ.							
Quercus nerifolia, Al. Br	R		Id				Id
Quercus stramineo, Lx	Col						
Quercus chlorophylla, Ung	R., Col., S. C						Id
Quercus Godeti, Hr	S. C						Id
Quercus Cleburni, Lx	B. B.						Rel
Quercus fraxinifolia, Lx	S. C						Rel
Quercus Ellisiana, Lx	F. E						Rel
Quercus Pealeii, Lx	F. E.						Rel
Quercus viburnifolia, Lx	Col., D B						Rel
Quercus angustiloba, Al. Br	Col						Id
Dryophyllum (Quercus) crenatum, Lx	P. of R						
Dryophyllum (Quercus) subfalcatum, Lx	P. of R					Rel	
Salix integra, Goepp	B. B.						Id
Populus melanaria, Hr	P. of R						Id
Populus melanarioides, Lx	P. of R						
Populus Ungeri, Lx	Col						
Populus mutabilis, var. ovalis, Hr	S. C., B. B			Id			Id
Populus monodon, Lx	R	Id					
Platanus Raynoldsii, Ny	Col., B. B						
Platanus rhomboidea, Lx	Col						
Platanus Haydenii, Ny	Col., B. B						
MOREÆ.							
Ficus irregularis, Lx	Col					Rel	
Ficus uncata, Lx	R., Col		Id				
Ficus Haydenii, Lx	B. B.						
Ficus Dalmatica, Ett	P. of R						Id
Ficus spectabilis, Lx	Col						
Ficus Smithsoniana, Lx	R						

Table of Distribution of the Species of the Laramie Group—Continued.

NAMES OF SPECIES.	Eaton Mountains, Placifxa, Colorado, Henry's Fork, Barrell's Springs, Fort Ellis, Spring Cañon, Black buttes, Point of Rocks, Yellowstone Lake.	Mississippi Eocene.	Green River Group—Oligocene.	Carbon, Alaska, etc.—Miocene.	Sézanne—Eocene.	Bornstädt, Mt. Promina—Oligocene.	Miocene.
Ficus occidentalis, Lx	Col						
Ficus planicostata, Lx	B. B., Col., P. of R						
Ficus planicostata, var. latifolia, Lx	Col						
Ficus planicostata, var. Goldiana, Lx	Col				Rel		
Ficus tiliæfolia, Al. Br	S.C.,P.of R.,R.,Col.,B.B		Id	Id			Id
Ficus subtruncata, Lx	Col						Rel
Ficus auriculata, Lx	Col., Sp. C						
Ficus asarifolia, Ett	Col., P. of R					Id	
NYCTAGINEÆ.							
Pisonia racemosa, Lx	B. B						Rel
LAURINEÆ.							
Laurus ocoteoides, Lx	Col						
Laurus prætans, Lx	P. of R	Rel					Rel
Cinnamomum affine, Lx	Col	Id	Id	Id			Rel
Cinnamomum Scheuchzeri, Hr	Sp. C						Id
Cinnamomum polymorphum, Hr	Col						Id
Daphnogene Anglica?, Hr	Col						
LONICEREÆ.							
Viburnum marginatum, Lx	B. B., P. of R., Col				Rel		
Viburnum platanoides, Lx	B. B				Rel		
Viburnum rotundifolium, Lx	B. B., P. of R				Rel		
Viburnum dichotomium, Lx	B. B				Rel		
Viburnum Whymperi, Hr	B. B., P. of R						Id
Viburnum Lakesii, Lx	Col						
Viburnum anceps, Lx	Col						
Viburnum Goldianum, Lx	Col						
Viburnum solitarium, Lx	Col						
OLEACEÆ.							
Fraxinus Goldiana, Lx	Col						
Fraxinus eocenica, Lx	Col						
Fraxinus denticulata, Hr	Sp. C						Id
DIOSPYRINEÆ.							
Diospyros? ficoidea, Lx	B. B						
Diospyros brachysepala, Al. Br	Col						Id
ERICACEÆ.							
Andromeda Grayana, Hr	Sp. C			Id			
ARALIACEÆ.							
Aralia pungens, Lx	Col				Rel		
Cissus laevigata, Lx	Col						
Cissus lobato-crenata, Lx	Col., B. B						Rel
Cissus tricuspidata, Hr	B. B						Id
Vitis Olriki, Hr	R						Id
Vitis sparsa, Lx	B. B						

Table of Distribution of the Species of the Laramie Group—Continued.

NAMES OF SPECIES.	Raton Mountains, Placker, Colorado, Henry's Ford, Bar-rell's springs, Fort Ellis, Spring Cañon, Black Buttes, Point of Rocks, Yellowstone Lake.	Mississippi Eocene.	Green River Group—Oligocene.	Carbon, Alaska, etc.—Miocene.	Sézanne—Eocene.	Bornstadt, Mt. Promina—Oligocene.	Miocene.
CORNEÆ.							
Cornus suborbifera, Lx	Col						Rel
Cornus Studeri, Hr	Col						Id
Nyssa lanceolata, Lx	Col, Sp. C						Rel
MAGNOLIACEÆ.							
Magnolia Lesleyana, Lx	R., Col	Id					
Magnolia tenuinervis, Lx	Col., B. B						
Magnolia Inglefieldi, Hr	Col						Id
Magnolia Hilgardiana, Lx	R	Id					
Magnolia attenuata, Lx	R						Rel
ANONACEÆ.							
Anona robusta, Lx	Col						
NYMPHEACEÆ.							
Nelumbium Lakesii, Lx	Col						
Nelumbium tenuifolium, Lx	Col						
MALVACEÆ.							
Sterculia modesta, Sap	Col				Id		
BÜTTNERIACEÆ.							
Dombeyopsis plataniodes, Lx	Sp. C., F. E.						Rel
Dombeyopsis trivialis, Lx	Col						
Dombeyopsis obtusa, Lx	Col						
Dombeyopsis grandifolia, Ung	Col				Id		
TILIACEÆ.							
Greviopsis Saportana, Lx	B. B				Rel		
Greviopsis tenuifolia, Lx	B. B				Rel		
Greviopsis Cleburni, Lx	P. of R				Rel		
Apeibopsis discolor, Lx	B. B						Rel
SAPINDACEÆ.							
Sapindus caudatus, Lx	Col., B. B.						Rel
CELASTRACEÆ.							
Celastrinites artocarpoides, Lx	Col					Rel	
Celastrinites lævigatus, Lx	Sp. C					Rel	
RHAMNEÆ.							
Paliurus zizyphoides, Lx	Col., B. B						
Zizyphus distortus, Lx	Col						
Zizyphus Beckwithii, Lx	Col						
Zizyphus fibrillosus, Lx	Col., B. B						
Berchemia multinervis, A. Br	R						Id
Rhamnus alaternoides, Hr	Col						Id
Rhamnus rectinervis, Hr	Col., B. B						Id
Rhamnus inæqualis, Lx	Col						Rel
Rhamnus discolor, Lx	B. B						

Table of Distribution of the Species of the Laramie Group—Continued.

NAMES OF SPECIES.	Raton Mountains, Placiere, Colorado, Henry's Fork, Barrell's Spring, Fort Ellis, Spring Cañon, Black Butter, Point of Rocks, Yellowstone Lake.	Mississippi Eocene.	Green River Group—Oligocene.	Carbon, Alaska, etc.—Miocene.	Sézanne—Eocene.	Dornstadt, Mt. Promina—Oligocene.	Miocene.
			AMERICAN.			EUROPEAN.	
Rhamnus Cleburni, Lx	Col., D. B				Rel		
Rhamnus Goldianus, Lx	Col., D. B			Id	Rel		
Rhamnus obovatus, Lx	R. Col						
Rhamnus salicifolius, Lx	Col., B. D	Rel					
Rhamnus deformatus, Lx	Col						
Rhamnus Rossmässleri, Ung	B. D						Id
JUGLANDEÆ.							
Juglans rhamnoides, Lx	S. C., B. D., P. of R						Rel
Juglans Lecouteana, Lx	Col						Rel
Juglans rugosa, Lx	S. C., D. B., Col			Id			Rel
Juglans thermalis, Lx	Col						
Juglans Schimperi, Lx	Col		Id				
ANACARDIACEÆ.							
Rhus membranacea, Lx	P. of R						Rel
Rhus pseudo-Meriani, Lx	B. B						Rel
HALORAGEÆ.							
Trapa microphylla, Lx	P. of R						
MYRTACEÆ.							
Eucalyptus Hæringiana?, Ett	B. D					Id?	
LEGUMINOSÆ.							
Podogonium Americanum, Lx	B. B		Id				
Leguminosites cassioides, Lx	S. C		Id				
INCERTÆ SEDIS.							
Carpites oviformis, Lx	Col						
Carpites triangulosus, Lx	Col., P. of R						
Carpites costatus, Lx	Col						
Carpites cassæformis, Lx	Col						
Carpites myricarum, Lx	B. B						
Carpites rostellatus, Lx	Col						
Carpites mitratus, Lx	B. D						
Carpites verrucosus, Lx	B. B						
Carpites minutulus, Lx	Col						
Carpites viburni, Lx	B. D						Rel
Carpites spiralis, Lx	Pl						
Carpites rhomboidalis, Lx	Col						
Carpites bursæformis, Lx	B. B						
Carpites ligatus, Lx	Pl						
Carpites valvatus, Lx	B. B						

DESCRIPTION OF SPECIES ADDED TO THE FLORA OF THE LARAMIE GROUP.

FILICES.

Osmunda major, sp. nov.

Plate XVIII. Fig. 5.

Frond pinnate; pinnules simple, alternate, large and thick, linear-lanceolate, unequilateral at base; borders undulate; medial nerve narrow; lateral nerves passing to the borders at a broad angle of divergence, forking generally once from the base, one of the branches sometimes forking again from the middle.

This beautiful fragment seems to belong to the same species as that of fig. 5, pl. iv, "U. S. Geol. Rep.," vii; at least the nervation is identical in its characters. The borders of the leaflets, however, are very entire, while they are obscurely crenulate in pl. iv, fig. 5. They come from the same locality. On the other hand the fragments, figs. 6 and 7 of pl. iv, vol. vii, have the same nervation as fig. 1—that is, a very narrow midrib, and the lateral veins forking more generally from the middle than from the base. It is, therefore, uncertain whether these fragments represent two or three species, or whether, perhaps, they may all be referable to the same.

Hab.—Golden. *A. Lakes.* Collection of Princeton College.

Pteris erosa, Lesqx.

Plate XIX, Fig. 1.

"U. S. Geol. Rep.," vii, p. 53, pl. iv, fig. 8.

Fronds simply pinnate; pinnæ large, linear-oblong, narrowed to a pointed acumen, unequilateral at base; lateral nerves distant, obtusely diverging from the medial nerve, curving down in joining it, forking at the base only, rarely one of the veins forking again from the middle.

By the shape of its leaflets and their nervation this species resembles the former and should, perhaps, be identified with it. The borders are sharply irregularly serrate, sometimes merely gnawed in places.

Hab.—Same locality as the preceding; also communicated by Mr. *Lakes.* It is the property of the Princeton College.

Gymnogramma Haydenii, Losqx.

Plate XIX, Fig. 2.

"U. S. Geol. Rep.," vii, p. 59, pl. v, figs. 1-3.

The fragment represented here is the upper part of a large leaflet having exactly the same specific characters. It has been figured, on account of the locality, as a positive identification of Snake River and Yellowstone Lake with the Laramie Group.

Hab.—Golden. *A. Lakes.*

PALMÆ.

Oreodoxites plicatus, sp. nov.

Plate XVIII, Figs. 1-4.

Leaves acute at both ends, deeply plicate lengthwise in numerous rays converging at the base and the apex, obscurely marked toward the base by a narrow medial nerve; rays distinctly veined; primary nerves distinct, separated by 3 or 4 thin intermediate ones.

On account of the plicate lamina, the leaves are referable to palms, and, as seen by figs. 2 and 3, they appear partly traversed by a narrow rib, which would indicate the disposition of the leaves as simple: but they are more probably lobes of a compound or palmately divided frond, like those of *Oreodoxia regia* of Cuba. In this last species the lobes are much longer and comparatively narrower, connected near the base. This disposition may have been the same for the fossil leaves, as the fragments, figs. 2 and 3, appear as lacerated near the base, and therefore as if they had been merely segments of a palmately divided frond.

The fragments of leaves described as *Ludoriopsis Geonomæfolia*, Sap., "Fl. de Sézanne," p. 339, pl. iv, fig. 1, are the only fossil plants to which the species might be compared. If the midrib of fig. 2 was more distinctly marked and the rays flat, the likeness would be striking. Saporta's species is referable to the *Pandaneæ*. It has not the truly plicate rays of the palms.

Hab.—Golden, Colorado. Found by Rev. *A. Lakes.* The specimens belong to the Museum of Princeton College.

OLEACEÆ.

Fraxinus coccinea, Lesqx.

Plate XX, Figs. 1-3.

"U. S. Geol. Rep.," vii, p. 229.

This fine species has been fully described, as quoted above. The specimens which represent it belong to the Princeton Museum.

ARALIACEÆ.

Aralia pungens, sp. nov.

Plate XIX, Figs. 3, 4.

Leaves coriaceous, rigid, very large, palmately divided; segments deeply cut into lanceolate sharply acuminate lobes—the lower opposite, the upper simple or lobate on one side.

The general outline of the leaves represented by the figured fragments is very probably analogous to the one figured in pl. xxxv of this volume; for it is evident that we have here mere segments or fragments of a compound leaf. These segments are subdivided into long lanceolate sharply acuminate entire lobes, which, oblique at their base, are turned up and erect at the apex. The nervation of the segments is pinnate; the lower secondary veins are opposite, strong, passing up to the point of the lobes, or curving up and following close to the borders like the lateral veins of the lobes.

This species is allied in its form to what has been described in vol. vii as *M. Lessigii*, p. 138, but the nervation differs. In *M. Lessigii*, the tertiary veins directed toward the sinuses divide under them into two branches, passing along on both sides and following the borders of the lobes, while in this leaf the tertiary veins do not divide, but appear to merely pass up on one side without forking. Though this difference may be marked, it is scarcely possible to doubt that these fragments represent the same group or the same genus of plants, and, as I have remarked it in the description of *M. Lessigii*, Saporta and other authors refer plants of this kind to the *Araliaceæ*.

The fossil leaves, published thus far, and more evidently related to these fragments, are the species of *Sylphidium*, Massalongo, on which Schimper remarks that the three species described from fragments are without doubt referable to the genus *Aralia* and represent a single species, perhaps identical with *Aralia multifida*, Sap.

Hab.—Golden. *A. Lakes.* Specimens in the Museum of Princeton College.

MAGNOLIACEÆ.

Magnolia tenuinervis, Lesqx.

Plate XIX, Fig. 6.

"U. S. Geol. Rep.," vii, p. 249, pl. xlv, figs. 1–5.

In the description of the species, *l. c.*, I compared the fragments by which it is represented to *M. Inglefieldi*, Heer, "Fl. Arct.," p. 120, especially to figs. 1–3 of pl. xviii. The part of leaf now figured is exactly of the same form as fig. 1 of this last plate. It is coriaceous, the surface smooth or glossy, the lateral veins only being apparently not quite as strong. The relation is therefore so close that it is scarcely possible to admit the difference as specific, the more so as some of the leaves figured in vol. vii have the lateral nerves quite as strong as represented by Heer.

Hab.—Golden. *A. Lakes.* Specimen in the National Museum.

ANONACEÆ.

Anona robusta, sp. nov.

Plate XX, Fig. 4.

Leaves large, coriaceous, ovate-lanceolate, gradually narrowed to the pointed apex, rounded at base, pinnately nerved; secondary nerves strong, close, parallel, curved in passing to the borders, camptodrome.

The leaf is about 13 centimeters long, 6 broad below the middle; the borders are slightly undulate; the medial nerve is thick; the lateral (12 pairs) also thick, especially toward the base, are alternate, very open or nearly at right angles toward the base, then gradually at a more acute angle of divergence, which in the upper ones is only 30°. These veins are all simple, more or less obliquely cut by strong nervilles, which are either simple and continuous or anastomosing in the middle of the areas.

The species is distantly related to *Anona elliptica*, Ung., "Syllog.," iii, p. 43, pl. xiv, fig. 1. The nerves, however, are much stronger indeed stronger than in any fossil leaf referred to this genus, and the base of the leaf is rounded.

Hab.—Golden, Colorado. Rev. *A. Lakes.*

STERCULIACEÆ.

Sterculia modesta, Sap.

Plate XX, Fig. 5.

Leaves thick, rounded in the lower part, trilobate at the apex; medial lobe longer, separated from the lateral by broad sinuses; nervation trifid from the base; lateral nerves camptodrome.

This finely preserved leaf is 8 centimeters long from the base to the apex of the middle lobe, and 6 centimeters broad between the points of the lateral ones. It is enlarged in the middle, a little contracted below the lateral lobes, and deltoid to the apex. The primary nerves are strong; the lateral are entwined by distinct nervilles; the areolation is in loose irregularly quadrate meshes.

By comparison with a fragment described under this name in "Fl. de Sézanne," p. 401, pl. xii, fig. 2, the American leaf has been identified by the author.

Hab.—Golden, Colorado. *A. Lakes.* Specimen in the Museum of Princeton College.

FRANGULACEÆ.

Zizyphus Beckwithii, sp. nov.

Plate XIX, Fig. 5.

Leaf membranaceous, oval or obovate, rounded at the top, narrowed and decurrent to the petiole, palmately tri-nerved from the base; medial nerve narrow, with a single branch above the middle, the lateral curving up at a distance from the borders nearly acrodrome, much branched outside; nervilles close, distinct, at right angles to the midrib.

The fine leaf, somewhat fan-like, 4½ centimeters long, 3 broad, has a thick petiole a little more than 1 centimeter long. The lateral primary nerves ascend to the top at equal distances from the midrib and the borders, which are perfectly entire. The secondary nerves are numerous (about 12 pairs), parallel, the lower being basilar and marginal; the nervilles are strong, parallel, continuous, and very close. The species is related to *Zizyphus Raincourti*, Sap., of the Sézanne flora.

Hab.—Near Golden, Colorado. *H. C. Beckwith.* Specimen in the National Museum.

Rhamnus deformatus, sp. nov.

Plate XX, Fig. 6.

Leaf lanceolate, tapering to an obtuse point, abruptly narrowing and decurrent to the petiole; borders entire, irregularly undulate; lateral nerves simple, campto-drome.

The leaf seems to have been deformed in the process of maceration. It is largest below the middle, diversely undulate-plicate on both sides; the secondary nerves are numerous (16 pairs), open, but much curved in passing toward the borders and following close to them, the upper ones at a more acute angle of divergence than those of the base.

Hab.—Golden, Colorado.   Specimen in the National Museum.

THE FLORA OF THE GREEN RIVER GROUP.

GEOLOGICAL DISTRIBUTION OF THE MEASURES.

In my preceding Reports I have referred to the Green River Group a limited number of species of fossil plants obtained from different localities mentioned below, and which were formerly considered as pertaining to the same geological stage. Now this group includes four members: the lower, the Wasatch, of which the Green River is an upper member; then, in ascending, the Bridger, the Uinta, and the White River with the Oregon beds.

The name of the Green River Group was proposed by Dr. F. V. Hayden on account of the great extent, thickness, and display of strata of this formation along Green River in Wyoming.

The formation as it is seen there is purely of a fresh-water origin and seems to be a continuation of the Eocene Laramie Group, or Lignitic, its strata being conformable to it and the modifications of the compounds being gradual. The lower member of the measures is mostly composed of arenaceous beds, the upper a series of laminated shale, each of these members averaging about one thousand feet in thickness.

The upper part of the measures merit especially to be considered now, as from it are derived the fossil remains which have been described here as derived from the Green River Group.

The shale, variegated in color, mostly red and white, and variable in thickness, give to the measures a peculiar banded appearance, especially marked near Green River Station, where I had an opportunity to make some observations on the distribution of the strata. At this place a section of 550 feet from the bed of the river to the high round bluff towering there over the country around shows the multiplicity of the layers and the variety of the compound.[1] The upper part of the bluff is a hard ferru-

---

[1] Hayden's "Annual Report," 1872, p. 336, where the section is given in detail.

ginous red sandstone in layers varying from 6 inches to 1 foot; below this there are 55 feet of laminated argillaceous sandstone with remains of fishes and plants intercalated between distinct slaty layers ¼ to 1 inch thick; then five beds of black bituminous compact shale measuring 2, 5, 25 feet, separated by beds of white calcareous shale, sandstone in thin layers, etc. Few of the beds are compact and homogeneous except the bituminous shale. The intercalated sandstone, four beds, variable from 5 to 13 feet, are composed of shaly layers. Near the base of the section only there is a bed of hard calcareous somewhat compact rock, which I have not remarked elsewhere in the country around.

The localities where fossil plants formerly referred to the Green River Group have been obtained are near Alkali Stage Station and Green River Station, Wyoming; in Randolph County of the same State; near Elko Station, on the U. P. Railroad, in Nevada; near the mouth of White River, Utah; and especially at Florissant, a locality also mentioned as Castello's Ranch and South Park, in Colorado.

The beds[1] of Florissant, now generally known for the abundance of their fossil remains, plants and insects especially, have been formed by like deposits. The geologist, Dr. A. C. Peale, one of the assistants of Dr. F. V. Hayden in his Survey of the Territories, has first given a short account of the formation near Florissant, a settlement rather than a village, situated in a narrow valley of the mountains, at the southern extremity of the Front Range of Colorado. He says: "In this valley, the name of Hayden Park has been given to the low rolling country to the west of Pike's Peak. Hayden Park is drained by Front Creek, West Creek, and Beaver Creek. The latter flows to the northwest and empties into the South Platte just below the upper cañon. About five miles from its mouth, around the settlement of Florissant, is an irregular basin filled with modern deposits. The entire basin is not more than five miles in diameter. The deposits extend up the branches of the creek, which all unite near Florissant. Between the branches are granite islands appearing above the beds which themselves rest on the granite. Just below Florissant, on the north side of the road, are bluffs not over 50 feet in height,

---

[1] Dr. Hayden's "Annual Report, U. S. Geological Survey of the Territories," 1873, p. 200.

in which are good exposures of the various beds. The following section gives them from top downward:

"1. Coarse conglomerated sandstone.

"2. Fine-grained, soft, yellowish-white sandstone, more or less argillaceous, and containing fragments of stems and leaves.

"3. Coarse gray and yellow sandstone.

"4. Chocolate-colored clay shales with fossil leaves. At the upper part the shales are black, and below pass into—

"5. Whitish clay shales.

"These last form the base of the hill. The beds are all horizontal."

After remarking on the presence of fragments of trachyte scattered around and found in layers near the surface, as seen by the boring of a well in the vicinity, Dr. Peale continues: "The lake basin may possibly be one of a chain of lakes that extended southward. I had thought it possible that the beds were of Pliocene age. The specimens obtained from No. 4 of the section above were submitted to Mr. Lesquereux, who informs me that they are Upper Tertiary, and says that he does not believe, as yet, that the plants of the Green River Group, to which are referable the specimens sent to him, authorize the conclusion of Pliocene age. He rather considers them, as yet, as Upper Miocene. The species known of our Upper Tertiary are, as yet, too few and represented in too poor specimens for definitive conclusion. Those sent from Florissant have a *Myrica*, a *Cassia*, fragments of leaves of *Salix angustata*, Al. Br., a *Rhus*, an *Ulmus*, and a fragment of *Poa* or *Poacites*."

I give the end of the quotation in order to show that the first opinion I expressed on the age of the Green River Group from its vegetable remains was based upon the examination of too insufficient materials.

After Dr. Peale the lake basin of Florissant has been carefully explored by Professor Sam. H. Scudder, who, in "Bulletin of the Geol. Survey," vol. vi. No. 2, has given in great detail the most precise and interesting account of his researches. It comprises not only the topographical description of the basin, the geology and stratigraphy of the beds formed by deposits of the lake, but a preliminary report on the insects and the plants obtained there by himself in an immense number

C F 9

of specimens. From this valuable memoir are derived a few notes which complete what the paleontologist may wish to know in regard to the strata from which the fossil remains are derived.

Professor Scudder's memoir is elucidated by a map of the Tertiary basin of Florissant as it was at the time when the strata were deposited. The area was then covered by a shallow sheet of water, hemmed in on all sides by near granite hills whose wooded slopes come to the water's edge, sometimes, especially on the northern and eastern sides, rising abruptly; at others gradually sloping so that reeds and flags grew in the shallow water by the shore; the water of the lake, penetrated by deep inlets between the hills, giving to it a varied and tortuous outline. This old lake was really a long outlet following the bottom of the valley, and expanding on both sides in lateral long shallow straits or pools. In one place the lake is contracted to half a mile in width; at two others one-fourth of a mile; taken altogether it is on an average 1 mile broad, being 6 to 7 miles long, expanding, on the eastern side especially, into nine of those narrow shallow straits. The outlines of the straits are, of course, varied. The area covered by their water measured half a mile to a mile long, one-fourth to half a mile broad, so that the shape of that Tertiary lake, as it is represented upon the map, resembles an oblong leaf, lobate on the borders, somewhat like a leaf of the white oak. It is easy to understand how those shallow pools, penetrating between hills covered with deep forest, alternately drying in summer and filling up in the rainy season, could become the reservoirs of woody and animal débris thrown upon their surface from overhanging trees and rocks, and there periodically accumulating by the succession of dryness and flood.

Professor Scudder supposes that the ancient outlet of the whole system was at the southern extremity; at least, the marks of the lake deposits reach near the ridge which now separates the waters of the Platte and of the Arkansas; and the nature of the basin itself, the much more rapid descent of the present surface on the southern side of the division, with the absence of any lacustrine deposits upon its slopes, lead to this conclusion.

Says Professor Scudder: "The very shales of the lake itself, in which the myriad of plants and insects are entombed, are wholly composed of volcanic sand and ash; 50 feet or more thick, they lie in alternating layers of coarser and finer materials. About half of this, now lying beneath the general surface of the ground, consists of heavily bedded drab shales with a conchoidal fracture, and totally destitute of fossils. The upper half has been eroded and carried away, leaving, however, the fragmentary remains of this great ash deposit clinging to the borders of the basin and surrounding the islands; a more convenient arrangement for the present explorer could not have been devised. That the source of volcanic ashes must have been close at hand seems abundantly proved by the difference in the deposits at the extreme ends of the lake. Not only does the thickness of the beds differ at the two points, but it is difficult to bring them into anything beyond the most general concordance.

"The excavation of the filled-up basin we must presume to be due to the ordinary agencies of atmospheric erosion. The islands in the lower lake take now as then the form of the granitic nucleus; nearly all are long and narrow, but their trend is in every direction, both across and along the valley in which they rest. Great masses of the shales still adhere equally on every side to the rocks against which they are deposited, proving that time alone, and no rude agency, has degraded the ancient flora of the lake."

The examination of Professor Scudder of the deposits of this lacustrine basin was principally made in a small hill, from which, perhaps, the largest number of fossils have been taken, lying just south of the house of Mr. Adam Hill and upon his ranch. "Like the other ancient islets of this upland lake it now forms a mesa, or flat-topped hill, about 30 to 50 feet high, perhaps 300 feet long and 80 broad. Around its eastern base are the famous petrified trees, huge, upright trunks, standing as they grew, which are reported to have been 18 to 20 feet high at the advent of the present residents of the region. Piecemeal they have been destroyed by vandal tourists, until now not one of them rises more than 2 or 3 feet above the surface of the ground, and many of them are entirely leveled; but their huge size is attested by the relics, the largest of which can be seen to have been 10 to 15 feet in diameter. These gigantic trees appear to be Sequoias, as far as can be told from thin sections of the wood submitted to Dr. L. Goodale. As is well known, remains of more than one species of *Sequoia* have been found in the shales at their base.

"From what information we could gain of the wells in this neighborhood, it would appear that the present bed of the ancient Florissant lake is entirely similar in composition for at least 30 feet below the surface, consisting of heavily bedded non-fossiliferous shales having conchoidal fracture. Above these basal deposits, on the slope of the hill, we found the following series from above downward, commencing with the evenly bedded strata:

"*Section in Southern Lake—By S. H. Scudder and A. Lakes.*

	Ft.	In.
1. Finely laminated, evenly bedded, light-gray shale; plants and insects scarce and poorly preserved	1	2
2. Light-brown, soft and pliable, fine-grained sandstone; unfossiliferous	2	0
3. Coarser, ferruginous sandstone; unfossiliferous	1	4
4. Resembling No. 1, leaves and insect remains	8	2
5. Hard, compact, grayish-black shale, breaking with a conchoidal fracture, seamed in the middle with a narrow strip of drab shale; fragments of plants	11	0
6. Ferruginous shale; unfossiliferous	5	0
7. Resembling No. 5, but having no conchoidal fracture; stems of plants, insects, and a small bivalve mollusk	3	4
8. Very fine gray ochreous shale; non-fossiliferous	0	2
9. Drab shales, interlaminated with finely divided paper shales of a light-gray color; stems of plants, reeds, insects	18	0
10. Crumbling ochreous shale; leaves abundant, insects rare	3	0
11. Drab shales; no fossils	3	0
12. Coarse ferruginous sandstone; no fossils	1	4
13. Very hard drab shales, having a conchoidal fracture and filled with nodules; unfossiliferous	24	7
14. Finely laminated yellowish or drab shales; leaves and fragments of plants, with a few insects	11	6

Ft. In.

15. Alternating layers of darker and lighter gray and brown ferru-
    ginous sandstone; no fossils 4 0
16. Drab shales; leaves, seeds, and other parts of plants, and in-
    sects, all in abundance................................ 24 0
17. Ferruginous, porous, sandy shale; no fossils 2 4
18. Dark-gray and yellow shales; leaves and other parts of plants.. 3 4
19. Interstratified shales, resembling Nos. 17 and 18; leaves and
    other parts of plants, with insects 7 0
20. Thickly bedded chocolate-colored shales; no fossils.............. 17 0
21. Porous yellow shale, interstratified with seams of very thin
    drab-colored shales; plants............................. 3 0
22. Heavily bedded chocolate-colored shales; no fossils............. 11 6
23. Thinly bedded drab shales; perfect leaves, with perfect and
    imperfect fragments of plants and a few broken insects....... 7 6
24. Thinly bedded light-drab shales, weathering, very light; without
    fossils; passing into... 7 6
25. Thick-bedded drab shales, breaking with a conchoidal fracture;
    also destitute of fossils.... .... 7 0
26. Coarse arenaceous shale; unfossiliferous........................... 3 4
27. Gray sandstone, containing decomposing fragments of some
    white mineral, perhaps calcite; no fossils...................... 70 0
28. Coarse, ferruginous, friable sandstone, with concretions of a
    softer material; fragments of stems.............................. 23 0
29. Thinly bedded drab shales, having a conchoidal fracture; some-
    what lignitic, with fragments of roots, etc...................... 10 0
30. Dark chocolate shales, containing yellowish concretions; filled
    with stems and roots of plants.................................. 10 0
Total thickness of evenly-bedded shales (D. of Dr. Wadsworth's
    note) above floor deposits.. 23 0

"The bed which has been most worked for insects and leaves, and in
which they are unquestionably the most abundant and best preserved, is
the thick bed, No. 16, lying half way up the hill, and composed of rapidly
alternating beds of variously-colored drab shales. Below this, insects
were plentiful only in No. 19, and above it in Nos. 7 and 9; in other beds

they occurred only rarely or in fragments. Plants were always abundant where insects were found, but also occurred in many strata where insects were either not discovered—such as beds 18 and 21 in the lower half and bed 6 in the upper half—or were rare, as in beds 10 and 14 above the middle and bed 23 below; the coarser lignites occurred only near the base.

"The thickest unfossiliferous beds, Nos. 20 and 27, were almost uniform in character throughout, and did not readily split into laminæ, indicating an enormous shower of ashes or a mud-flow at the time of their deposition; their character was similar to that of the floor-beds of the basin.

"These beds of shale vary in color from yellow to dark brown. Above them all lay, as already stated, from 4 to 6 feet of coarser more granulated sediments, all but the lower bed broken up and greatly contorted. These reached almost to the summit of the mesa, which was strewn with granitic gravel and a few pebbles of lava."

The specimens of Florissant representing the plants described in this memoir were mostly obtained by Professor Scudder, who had opportunity to purchase for Dr. Hayden a collection made by Mrs. Charlotte Hill, the proprietress of the land where are exposed the banks containing the richest fossiliferous shale. A little later a scientific exploration for the College of Princeton visited the same locality and obtained there also a great number of specimens; some of these, very fine, which were loaned me for examination, have been figured and described in this report. I have been allowed to use the names of some of the members of the exploration—Messrs. W. B. Scott, H. F. Osborn, F. Speir, McCosh, W. Libbey—for the nomenclature of some of the new species which are represented by the Princeton specimens.

ENUMERATION AND DESCRIPTION OF THE SPECIES OF FOSSIL PLANTS KNOWN FROM THE GREEN RIVER GROUP.

CRYPTOGAMÆ.

FUNGI.

Sphaeria myricæ, Lesqx.

"U. S. Geol. Rep.," vii, p. 34, pl. ii, fig. 4.

CHARACEÆ.

CHARA, Waill.

Chara? glomerata, sp. nov.

Plate XXI, Fig. 12.

Leaves short, in compact, dense, distant or terminal capitules; stem narrow.

These fragments are not positively referable to *Chara* on account of the compactness and shortness of the leaves. The branches bearing the capitules are smooth, flexuous, the leaves ? apparently subcylindrical, acute. They may represent flower-bearing pedicels of *Platanus* like *P. racemosa*, Nutt. They, however, can scarcely be considered as such, for not the least fragment of *Platanus* leaves has been found as yet in the Green River Group.

Hab.—Florissant. U. S. Geol. Expl. Dr. *F. V. Hayden.*

MUSCI.

FONTINALIS, Linn.

Fontinalis pristina, sp. nov.

Plate XXI, Fig. 9.

Leaves obscurely two ranked, crowded, linear-lanceolate, acuminate, ecostate.

The leaves are close, gradually enlarged toward the embracing base, about one centimeter long, very narrow.

Hab.—Florissant, Colorado. The locality indicated as Castello's ranch is the same.

135

HYPNUM, Linn.

Hypnum Haydenii, Lesqx.

"U. S. Geol. Rep.," vii, p. 44, pl. v, figs. 14-14b.

RHIZOCARPEÆ.

SALVINIA, Mich.

Salvinia cyclophylla, Lesqx.

"U. S. Geol. Rep.," vii, p. 64, pl. v, figs. 10, 10a.

Salvinia Alleni, Lesqx.

Plate XXI, Figs. 10, 11.

"U. S. Geol. Rep.," vii, p. 65, pl. v, fig. 11.

The species is common and has been obtained in large well-preserved specimens by the different collectors. The leaves are merely variable in size, obtuse or slightly emarginate at the apex, topped by the point of the excurrent nerve.

EQUISETACEÆ.

EQUISETUM, Linn.

Equisetum Wyomingense, Lesqx.

"U. S. Geol. Rep.," vii, p. 69, pl. vi, figs. 8-11.

Equisetum Haydenii, Lesqx.

"U. S. Geol. Rep.," vii, p. 67, pl. vi, figs. 2-4.

ISOETEÆ.

ISOETES ?, Web.

Isoetes brevifolius, sp. nov.

Tufts small, compact; leaves cylindrical, acuminate, coming out of a small cylindrical stem or rhizoma.

The leaves are 1 to 2 millimeters in diameter, 4 to 6 centimeters long, narrowed to a point, apparently smooth. The small tufts much resemble *Isoetes Braunii*, Heer, as figured in "Fl. Tert. Helv.," pl. xiv, fig. 5, the leaves being only shorter and narrower.

Hab.—Florissant. Specimen No. 66 of the collection of Mr. R. D. Lacoe, of Pittston, Penna.

LYCOPODIACEÆ.

LYCOPODIUM, Linn.

Lycopodium prominens, Lesqx.

"U. S. Geol. Rep.," vii, p. 45, pl. v, figs. 13–13*b*.

FILICES.

SPHENOPTERIS, Phill.

Sphenopteris Guyottii, sp. nov.

Plate XXI, Figs. 1–7.

Ultimate pinnæ linear-lanceolate, of various lengths; rachis narrow and narrowly winged by the decurrent base of the lanceolate obtuse pinnules; lower pinnules regularly divided into 2 to 4 half-round short lobes, connate in the middle; upper pinnules entire, oblong, obtuse; medial nerve thin, pinnately branching into oblique lateral nerves, generally forking once, rarely simple; substance of the leaves rather thin: nervation distinct.

This fern, common at Florissant, but always found in small fragments, has no near relation to any fossil species known to me, being only comparable to *Sphenopteris Blomstrandi*, Heer, " Fl. Arct." i, p. 155. pl. xxix. figs. 1–5, from the Miocene of Spitzbergen. In its form and its nervation it is a true *Phegopteris*, closely related to some Cuban species. *P. sericea*, *P. divergens*, &c. But from the absence of fructification an exact comparison is not possible.

Hab.—Florissant.	Seen in most of the collections.

ADIANTITES, Auct.

Adiantites gracillimus, sp. nov.

Plate XXI, Fig. 8.

Rachis very slender, filiform, flexuous, bearing at its top a few simple entire pinnules, oval in outline, sessile by the cuneate base, obtuse; nervation dichotomous, the medial nerves forking two or three times; branches very oblique, forking near the apex.

I have seen only the small fragment figured, which is, however, distinctly preserved. By the disposition of the leaflets and their shape it may be compared to *Asplenites allosuroides*, Ung., " Fl. v. Sotzka," which has small fructified pinnules; but the nervation is that of *Adiantum*.

Hab.—Florissant.

LASTRÆA, Presl.

Lastræa (Goniopteris) intermedia, Lesqx.

"U. S. Geol. Rep.," vii. p. 56, pl. iv, fig. 11.

PTERIS, Linn.

Pteris pseudo-pennæformis, Lesqx.

Ibid., p. 52, pl. iv, figs. 3, 4

DIPLAZIUM, Swartz.

Diplazium Muelleri, Heer.

Ibid., p. 55, pl. iv, figs. 10, 10a.

LYGODIUM, Sw.

Lygodium neuropteroides, Lesqx.

Ibid., p. 61, pl. v, figs. 4-7; vi, fig. 1.

Lygodium Dentoni, Lesqx.

Ibid., p. 63, pl. lxv, figs. 12, 13.

CONIFERÆ.

PINUS, Linn.

Pinus Florissanti, sp. nov.

Plate XXI, Fig. 13.

Strobile large, conical, 12 centimeters long or more, 6 centimeters in diameter at the broken base; scales large, 4½ centimeters long, 1½ broad; apophyses conical, transversely rhomboidal when flattened.

This fine cone is related to *Pinus ponderosa*, Douglas, a fine species of California and New Mexico, by the large size of the scales, not or scarcely enlarged under the apophyses.

Hab.—Florissant. U. S. Geol. Expl. Dr. *F. V. Hayden.*

Pinus palæostrobus ?, Ett.

"U. S. Geol. Rep.," vii, p. 83, pl. vii, figs. 25, 31.

SEQUOIA, Torr.

Sequoia angustifolia, Lesqx.

Ibid, p. 77, pl. vii, figs. 6-10.

Sequoia Langsdorfii, Brgt.

Ibid., p. 76.

Sequoia Heerii, Lesqx.

Ibid., p. 77, pl. vii, figs. 11-13.

Sequoia affinis, Lesqx.

Ibid., p. 75, pl. vii, figs. 3-5; lxv, figs. 1-4.

TAXODIUM, Rich.

Taxodium distichum miocenum, Heer.

"U. S. Geol. Rep.," vii, p. 73, pl. vi, figs. 12–14.
Abies Nevadensis, Lesqx., "Hayden's Ann. Rep.," 1872, p. 372.

WIDDRINGTONIA, Endl.

Widdringtonia linguæfolia, sp. nov.

Plate XXI, Figs 14, 14*a*.

Glyptostrobus Europæus, Lesqx., "U. S. Geol. Rep.," vii, p. 74, pl. vii, figs. 1, 2.

Branches and branchlets short, pinnately divided; divisions alternate; branchlets simple and slender; leaves appressed, irregularly two-ranked or subalternate, ovate, blunt-pointed or lingulate.

The specimens represent two forms of the same species, differing merely by the size or the thickness of the branches and branchlets. The more common form is figured; the other is more slender in all its parts, a var. *gracilis*, mentioned in "Hayden's Ann. Rep.," 1872, p. 371, as *Thuites callitrina*, Ung.

Hab.—Florissant. U. S. Geol. Expl. Dr. *F. V. Hayden.*

THUYA, Linn.

Thuya Garmani, Lesqx.

Hayden's "Ann. Rep.," 1872, p. 372.

GLYPTOSTROBUS, Endl.

Glyptostrobus Ungeri ?, Heer.

Plate XXII, Figs. 1–6*a*.

Heer, "Fl. Tert. Helv.," i, p. 52, pl. xviii; "Fl. Alask.," p. 22, pl. iii, figs. 10, 11.

Stem leaves squamiform, appressed, lanceolate, acute or acuminate; branch-leaves open, two-ranked, much longer, linear-lanceolate, acute; male cone small, oval, terminal; strobiles ovate on short branches; scales 6 to 9, obtusely dentate at the upper border, obscurely striate lengthwise.

This species, obtained in fine specimens, is in some of its characters identical with *Cupressites taxiformis*, Ung., "Chloris," p. 18, pls. viii and ix. The diversity of the leaves in regard to their position upon the stem and the base of the branches, where they are shorter, appressed, and squamiform, is not indicated by Unger. It seems also to be identical to *Chamæcyparites Hardtii*, Endl., as represented by Ett., "Häring Fl.," p. 35, pl. vi, figs. 1–21, two species referred by Schimper to *Sequoia Langsdorfii*, Brgt. The cones of the species of Florissant, however, are not those of a *Sequoia* but of a *Glyptostrobus*, and these, like the diversity in the form of the leaves, agree in character with *G. Ungeri*, Heer, quoted above, which is now considered by the author as a variety of *G. Europæus*. The cones only are somewhat larger, as figured by Heer, and the stem leaves rather obtuse than acuminate. As in the "Flora of Alaska," the same author represents these scaliform leaves acute, even acuminate, and as in that of Spitzbergen ("Fl. Arct.," iv, pl. xi, figs. 2–8) the same kind of leaves are either obtuse or acuminate, the reference of the American form to the species of Heer is sufficiently authorized. The species is closely related to *Glyptostrobus heterophyllus*, Endl., of China, the only living species of this genus.

Hab.—Very common at Florissant. The specimens figured are mostly those of the Princeton Museum.

PODOCARPUS, L'Hérit.

Podocarpus coccnica ?, Ung.

Leaves narrowly linear-lanceolate, acute, narrowed into a short petiole; medial nerve distinct.

This description refers to two leaves which agree with the description and figure of this species by Unger ("Fl. of Sotzka," p. 28, pl. ii, figs. 11–16). The medial nerve is flat and comparatively broad: the leaves are slightly broader in the middle.

Hab.—Florissant. No. 68 of Lacoe Collection.

GRAMINEÆ.

POACITES, Heer.

Poacites lævis, Heer.

Hayden's "Ann. Rep.," 1871, p. 285.

CYPERUS.

Cyperus Chavannesi, Heer.

"U. S. Geol. Rep.," vii, p. 92, pl. ix, figs. 1, 2.

CYPERITES, Lindl.

Cyperites Haydenii, sp. nov.

Plate XXIII, Figs. 1–3a.

Leaves large, gradually enlarging upward from its root, linear above; medial nerve broad and flat; lateral nerve parallel, distinct to the eye, separated by four or five very thin intermediate veins.

From the fragments preserved the leaves appear to have been very long. Linear in the middle where they are 3 centimeters broad, they are slightly narrower upward and apparently rounded to a pointed apex, gradually tapering downward to the upper part of the root, a small tubercle. The medial nerve, quite distinct, is 2 millimeters broad in the middle. Though related to *Cyperus* and *Cyperites*, this leaf has no marked affinity to any one of the numerous forms which have been described under this name. The leaf is quite flat and does not appear to have been keeled in the middle, but distinctly nerved. It comes out directly from the tubercle. The lateral nerves, 12 to 14, are separated by veinlets without any transverse veins.

Hab.—Randolph Co., Colorado. U. S. Geol. Expl. Dr. *F. V. Hayden.*

ARUNDO, Linn.

Arundo Goepperti ?, Münst.

" U. S. Geol. Rep.," vii, p. 86, pl. viii, figs. 3–5.

Arundo reperta, Lesqx.

Ibid., p. 87, pl. viii, figs. 6, 8.

PHRAGMITES, Trin.

Phragmites Alaskana, Heer.

Ibid., pl. viii, figs. 10-12.

TYPHACEÆ.

TYPHA.

Typha latissima, Al. Br.

Plate XXIII, Figs. 4, 4*a*.

Al. Br., "Stizenb. Verz.," p. 75; Heer, "Fl. Tert. Helv.," i, p. 98, pl. xliii, xliv; "Mioc. Balt. Fl.," p. 29, pl. iv, fig. 11; Ett., "Foss. Fl. v. Bilin," p. 30, pl. vi, fig. 9.

Leaves very long, 2 to 3 centimeters broad, linear, marked lengthwise by parallel strong nerves (14) crossed at right angles by transverse thin lines; intermedial veinlets numerous (10-13).

Though these fragments, which are numerous, and part of which only are figured, are referable to the European species by their appearance, they may represent a different one on account of the numerous intermediate veinlets which separate the primary nerves. In the European species only 4 to 6 are counted. while on the American specimens they are generally 10 to 12. It is. however. to be remarked that *Typha* species living at the present epoch have a wide range of distribution; the two species (*T. lati-folia* and *T. angustifolia*) are as common on the North American continent as they are in Europe.

Hab.—Florissant; Randolph County. U. S. Geol. Expl. Dr. *Hayden.*

POTAMOGETON, Linn.

Potamogeton? verticillatus, sp. nov.

Plate XXIII, Figs. 5, 6.

Stems slender; leaves verticillate or tufted, grass-like, linear-lanceolate, largest toward the base, sessile and narrowed to the point of attachment, nerved lengthwise in the middle; branches very slender, floating or pendant, bearing tufts of shorter leaves.

This species differs from its congeners by the position of the leaves in verticils upon apparently articulate stems. It is distantly related to *P. cæspitans*, Sap., "Ét.," i, p. 76, pl. iv, fig. 2.

Hab.—Florissant. The specimen (fig. 5) is from the Explor. of Dr. *F. V. Hayden;* the other belongs to the Princeton Museum.

Potamogeton geniculatus, Al. Br.

"Stizenb. Verz.," p. 75; Heer, "Fl. Tert. Helv.," i, p. 102, pl. xlvii, figs. 1-6; Ett., "Fl. v. Bilin," p. 29, pl. vii, figs. 1, 2.

Stems slender, branching, geniculate-flexuous; leaves narrowly linear, acuminate, fasciculate, sessile; fruits round or broadly oval-apiculate, 1 millimeter in diameter.

Though the specimens merely represent the upper part of a stem the characters of the leaves and the fructification refer the plant to Heer's species. The fruits are slightly smaller, however, rather round than ovate or exactly like those represented by the author, pl. xlvii, fig. 5c.

Hab.—Florissant. No. 69 of Lacoe Collection.

NAJADOPSIS, Heer.

Najadopsis rugulosa, sp. nov.

Plate XXIII, Fig. 7.

Stem dichotomous from inflated apicial innovations; segments flat, dichotomous, linear, acuminate, decurrent to the main stem; surface merely irregular and minutely wrinkled lengthwise, without trace of medial nerves.

The substance of this plant is somewhat thick; the leaves or segments seem to have been originally cylindrical, though quite flat upon the stone, by compression? All that can be seen of the plant is figured. It has an evident relation to *N. dichotoma*, Heer, "Fl. Tert. Helv.," i, p. 104, pl. xlviii, figs. 1–6. Not only the dichotomous disposition of the segments is analogous, but in fig. 1 of Heer the primary division appears as from an obscure innovation, while the top of the main stem seems to be inflated by the position of apparently fasciculate segments as they are in the middle of fig. 7 of our plate. The size of the European plant is smaller in all its parts.

Hab.—Florissant. U. S. Geol. Expl. Dr. *F. V. Hayden.*

MUSACEÆ.

MUSOPHYLLUM, Goepp.

Musophyllum complicatum, Lesqx.

"U. S. Geol. Rep.," vii, p. 96, pl. xv, figs. 1, 6.

The station of the bed of coal and shale where this plant was found in great profusion, with remains of *Sapindus obtusifolius*, appears rather referable to the Green River Group than to the Miocene of Carbon from the presence of this last species, which has been found also at Florissant.

AROIDEÆ.

ACORUS, Linn.

Acorus brachystachys, Heer.

"U. S. Geol. Rep.," vii, p. 105, pl. xiv, fig. 16.

LEMNACEÆ.

LEMNA, Linn.

Lemna penicillata, sp. nov.

Plate XXIII, Fig. 8.

Leaves small, round in outline, irregularly crenulate on the borders; surface rugose; rootlets numerous, in fascicles.

The leaves, 3 to 4 millimeters in diameter, are rugose on the surface and do not show any trace of nerves; they appear to have been fleshy, but they are quite flattened into thin flakes on soft shales.

Hab.—Florissant. U. S. Geol. Expl. Dr. *F. V. Hayden.*

PALMÆ.

FLABELLARIA, Schp.

Flabellaria Florissanti, sp. nov.

Plate XXIV, Figs. 1–2a.

Fronds large; rays diverging all around from the top of the nearly flat not keeled long rachis; rays large, very numerous, acutely keeled; primary nerves distinct; close intermediate veinlets, 3, 4.

This species has some degree of likeness to *Flabellaria cocenica*, Lesqx., "U. S. Geol. Rep.," vii, p. 3, pl. xiii, figs. 1–3. The rachis is not carinate but merely indistinctly lineate lengthwise, and the top of the rachis on one side of the leaf is also nearly truncate. The nerves are less distant and the intermediate veins less numerous. It is still more intimately related to *Flabellaria Lamanonis*, Brgt., and perhaps identical with it as figured in Sap., "Ét." i, p. 70, pl. iv. fig. 5,—at least the number of primary nerves in each division of the rays and that of the intermediate veins are about the same. The lateral rays are more sharply keeled in the American form and also more open, the lateral ones being at right angles to the more distinctly truncate top of the petiole.

Hab.—Randolph Co., Colorado. U. S. Geol. Expl. Dr. *F. V. Hayden.*

PALMOCARPON.

Palmocarpon? globosum, sp. nov.

Plate XXIV, Fig. 3.

Fruit large, globose, striate lengthwise.

The fruit is exactly globose, 18 millimeters in diameter; the testa appears to have been woody, though the fruit is flattened. This fruit has not been found in connection with the palm-leaf described above, but at a different locality, and therefore its reference to Palms is not positive. It resembles *Carpites lineatus*, Newby., as figured, pl. lx. fig. 1, "U. S. Geol. Rep.," vii, a species abundantly found at Evanston, where no remains of Palms have been discovered.

Hab.—Florissant. U. S. Geol. Expl. Dr. *F. V. Hayden.*

DICOTYLEDONES.

MYRICACEÆ.

MYRICA.

"U. S. Geol. Rep.," vii, p. 126.

§ 1. Leaves dentate, serrate or undulate.

Myrica Copeana, Lesqx.

Ibid., p. 131, pl. xvii, fig. 5.

Myrica obscura, sp. nov.

Plate XXXII, Figs. 8-10.

Leaves linear-lanceolate, coarsely serrate, rounded in narrowing to the petiole, unequilateral at base; nervation obsolete.

This form is related by its shape and the teeth of the borders to *M. Banksiæfolia*, Ung., as figured by Heer, "Fl. Tert. Helv.," pl. c, figs. 3-10, differing merely by the more rounded and unequilateral base of the leaves and the total disappearance of lateral nerves by immersion into a thick carbonaceous coating. However, fig 6 of Heer represents two leaves without traces of lateral nerves, and fig. 8 has the base somewhat rounded and unequilateral, though not quite as distinctly as in the American form. The pedicel of this last figure is also slender, of the same length as in fig. 10 of our plate. The leaves are on an average a little smaller than those of *M. Banksiæfolia*, 7 to 9 centimeters long and 1 to 2½ centimeters broad above the base; the teeth are generally sharp, slightly inclined upward.

Hab.—Florissant. U. S. Geol. Expl. Dr. *F. V. Hayden*.

Myrica Ludwigii, Schp.

"U. S. Geol. Rep.," vii, p. 133, pl. lxv, fig. 9.

Myrica acuminata, Ung.

Ibid., p. 130, pl. xvii, figs. 1-4.

Myrica rigida, sp. nov.

Plate XXV, Figs. 3, 4.

Leaves thick, rigid, subcoriaceous, lanceolate-acuminate, serrate, rounded and unequilateral at base, short petioled; medial nerve thin, straight, the lateral craspedodrome.

This species differs from the preceding by the distinctly lanceolate form of the leaves equally and gradually narrowing from the rounded base to the apex, by the short petiole, the distinct lateral veins and the

c F 10

blunt teeth of the borders. The leaves are also proportionally shorter, 5 to 7 centimeters long and 1 to 2 centimeters broad near the base. It is intermediate between the preceding and the following species.

Hab.—With the preceding.

Myrica Zachariensis, Sap.

Plate XXV, Fig. 5; XLVᵃ, Figs. 6–9.

Leaves very variable in size and shape, lanceolate and linear, narrowed and more or less decurrent to the petiole; medial nerve thick; lateral nerves open, curved in passing to the borders and along them; teeth entered by branchlets.

This species, as figured by Saporta, "Ét.," i, ii, p. 201, fig. 5, is represented in pl. xxv, fig. 5, and xlvᵃ, fig. 7. It is the variety *b. elongata.* The variety *c. angustifolia*, Sap., *loc. cit.*, fig. 1, has the character of pl. xlvᵃ, figs. 6–8, while fig. 9 of the same plate is exactly like a counterpart of fig. 10*b.*, Sap., "Ét.," ii, pl. 5, which is the variety *minuta* of this species. It differs from the two preceding species by the gradual narrowing of the base to the petiole, the border base being decurrent to it and bordering it to the point of attachment.

Hab.—Florissant. Specimens, pl. xlvᵃ, figs. 6–9, are from Alkali Station.

Myrica polymorpha, Schp.

Plate XXV, Figs. 1, 2.

Leaves thickish, membranaceous or subcoriaceous, long-lanceolate or linear-lanceolate acuminate, narrowed at base to a short petiole, serrate or dentienlate; primary nerves thick at base, the lateral more or less oblique, slightly curving in passing to the borders.

This species is described by Saporta as *Myricophyllum Zachariense*, "Ét.," i, ii, p. 220, pl. viii, fig. 2, with varieties *spinulosa* and *laciniata*, according to the more or less deep and acute teeth of the borders. Our plate represents the normal form. The leaves are long comparatively to their width—6 to 8 centimeters long, 5 to 6 millimeters broad. The species is, like the preceding, very polymorphous. The author compares it to the living *Myrica Æthiopica*, Linn., especially as to its nervation.

Hab.—Very common at Florissant.

Myrica callicomæfolia, sp. nov.

Plate XXVI, Figs. 5–14.

Callicoma microphylla, Ett., "U. S. Geol. Rep.," vii, p. 246, pl. xliii, figs. 2–4.

This species is evidently a *Myrica.* Better specimens show that the fragment which I considered as a compound leaf is a small branch with alternate leaves. The reference to *Callicoma* is not possible, as in this genus the divisions are opposite. Except from what is seen in the branch, fig. 5, whose divisions are alternate, distant, parallel, as well as the leaves, there is nothing to modify in the description of this species in vol. vii. *loc. cit.* The teeth are not always sharply acute, but more or less so, always inclined upward.

The species is closely related in the nervation to *M. Zachariensis*, var. *minuta*, Sap., *loc. cit.*, but differs evidently in the more rounded and unequilateral base of the leaves.

Hab.—Most abundant at Florissant, also at Elko Station, Utah.

Myrica fallax, sp. nov.
Plate XXXII, Figs. 11–16.

Very similar in its characters to the preceding species and perhaps a variety of it. It merely differs in the teeth being sharply acuminate or subspiniform, the lateral nerves less curved in passing toward the borders, the base of the leaves not as distinctly unequilateral. It is distantly related to *M. acuminata*, Ung.

Hab—Florissant. Not rare.

Myrica Scottii, sp. nov.
Plate XXXII, Figs. 17, 18.

Leaves coriaceous, long and narrow, linear-acuminate, narrowly cuneate to the petiole, sharply dentate; lateral veins more or less oblique and curved.

By the leaves, 6 to 9 centimeters long, 6 to 10 millimeters broad, with sharply spinescent teeth turned upward, the species is related to *M. Banksiæfolia*, Ung., and *M. obscura*, described above. It differs from both in the sharply dentate borders of the leaves, the lateral nerves being distinct and more acutely diverging.

Hab.—Florissant. Princeton Museum.

Myrica amygdalina, Sap.
Plate XXVI, figs. 1–4.
Sap., "Ét.," iii, ii, p. 21, pl. 1, figs. 8–10.

Leaves submembranaceous, oblong-lanceolate, obtuse or apiculate, narrowed to a short petiole, denticulate or subentire; secondary nerves numerous, at an acute angle of divergence, obliquely branching and reticulate.

The leaves are small, $2\frac{1}{2}$ to $5\frac{1}{2}$ centimeters long, enlarged toward the upper part; the areolation is distinct, formed by nervilles crossing the oblique divisions of the lateral nerves at right angles.

Hab.—Florissant. U. S. Geol. Expl. Dr. *F. V. Hayden.*

Myrica nigricans, Lesqx.

" U. S. Geol. Rep.," vii. p. 132, pl. xvii, figs. 9-12.

Myrica Bolanderi, Lesqx.

Ibid., p. 133, pl. xvii, fig. 17.

Myrica undulata, Lesqx.

Ibid., p. 131, pl. xvii, fig. 5.

Leaves lobate; lobes irregular, often serrate.

§2. Leaves pinnately lobed (*Comptonia*).

Myrica partita, Lesqx.

Ibid., p. 134, pl. xvii, fig. 14.

Myrica diversifolia, sp. nov.

Plate XXV. Figs. 6-15.

Leaves membranaceous, short-petioled, either longer, deeply lobate and lanceolate, or shorter, broadly ovate, diversely tri-quadri-lobate; lobes dentate; primary nerves narrow, the secondary open, curved in passing to the points of the lobes or of the teeth, branching; tertiary nerves in the direction of the sinuses, forking under them, each branch following the borders. Seeds small, oval-acute.

At first it is difficult to see that these leaves are referable to the genus *Myrica* and that they all represent the same species. In comparing, however, fig. 6 to *Myrica Graffii*, Heer. "Fl. Tert. Helv.," iii. p. 176, pl. cl, figs. 19, 20, the character of the nervation, the form of the leaves, the dentate lobes will be found much alike. The species are far different but the type is the same. The same degree of affinity is remarked between figs. 11-13 of our plate with *Myrica latiloba*, Heer. figs. 12-15 of the same plate; there is also a marked degree of relationship between the leaves I refer to this species and *Comptonia laciniata*, Ung.. "Fl. von Sotzka." p. 31, pl. viii, fig. 2.

Comparing now with one another the fragments which represent this species, we see in fig. 8 the same characters exactly as in fig. 6, merely modified by the shortening of the leaves and of their lobes. Fig. 11 represents an intermediate form, and with its deep-cut lobes fig. 13 is like an original representation of fig. 11. Indeed, considering the characters of these leaves with the more or less broadly cuneate base decurrent to the short petiole. their sharply dentate lobes, the membranaceous substance, the nervation. I am not able to find any difference to separate them into two or more species, and still less to refer them to a different genus. Some of the leaves (fig. 14 especially) have some of the characters of *Cratægus*, but the nervation recalls them to *Myrica*. The small seed, fig. 15, though a seed of *Myrica*, is not positively referable to this species.

Hab.—Florissant.   U. S. Geol. Expl. Dr. *F. V. Hayden.*

Myrica latiloba, Heer, var. acutiloba.

" U. S. Geol. Rep.," vii, p. 134. pl. xvii, fig. 13.

§ 3. Leaves pinnately lobed (*Comptonia*).

The leaf mentioned with the description of this species as being identical in character with it and obtained from the Miocene of Oregon is figured, pl. l. fig. 10, and described with Miocene plants.

Myrica Brongniarti ?, Ett.

"U. S. Geol. Rep.," vii, p. 135, pl. xvii, fig. 15.

Myrica Alkalina, sp nov.

Plate XLVa, Figs. 10–15.

Leaves short, trilobate and obtusely dentate from a cuneate base, or lanceolate, rounded and narrowed to the base, pinnately, obtusely or acutely dentate.

The species represented by a large number of fragmentary leaves, mixed upon the same specimens, present two forms, rather marked varieties, especially differing by acute or obtuse lobes or teeth. The leaves are subcoriaceous or membranaceous, somewhat large, 3 to 8 centimeters long, 2½ to 3 centimeters broad, either lobate with narrow cuneate base, or pinnately deeply dentate, more or less obtusely cuneate at base. The medial nerve is thick; the lateral nerves, at a broad angle of divergence, much curved in passing up to the points of the lobes, are generally separated by parallel shorter tertiary veins, anastomosing with oblique nervilles or branchlets derived from the secondary nerves.

The species is comparable to both *Myrica Vindobonensis*, Ett., in Heer, " Fl. Tert. Helv.," p. 34, pl. lxx, figs. 5, 6, and *M. Ungeri*, Heer, *l. c.*, p. 35, pl. lxx. figs. 7, 8, differing from both by shorter comparatively broader leaves, more equally dentate-lobed.

As represented upon the plates, the leaves would seem to be referable to two different species. The fragments, however, are so well mixed together that sometimes one leaf appears acutely dentate on one side and obtusely so on the other.

Hab.—Alkali Station, Wyoming. Professor *Scudder*.

Myrica insignis, Lesqx.

" U. S. Geol. Rep.," vii, p. 135, pl. lxv, figs. 7, 8.

This species has a degree of relationship to the preceding.

BETULACEÆ.

BETULA, Linn.

" U. S. Geol. Rep.," vii, p. 137.

Betula Florissanti, sp. nov.

Plate XXVII. Fig. 11.

Leaves small, lanceolate-acuminate, unequilateral at the cuneate base, borders doubly serrate; medial nerve thin; secondary nerves generally opposite, curved in passing to the borders, branching, entering the teeth like the branches and united by nervilles.

The leaf, 5½ centimeters long, 1½ broad, appears unequilateral at the narrowed base. The primary and secondary teeth are small, acute, and turned upward.

Hab.—Florissant. Princeton Museum.

Betula truncata, sp. nov.

Plate XXVIII. Figs. 7, 8.

Leaves short and short-petioled, ovate-lanceolate, truncate or rounded at base, simply dentate; lateral veins at a broad angle of divergence, numerous, parallel, the lower opposite.

The leaves, 3 to 4 centimeters long, 2 centimeters broad, equally dentate from near the base, have the secondary nerves at an angle of divergence of 60°, generally branching. The relation of this species is to *Betula crenata*, Ung., " Schoss. Fl.," p. 11, pl. iii, figs. 7, 8. The lateral nerves are more open, more numerous, and less curved in the American species.

Hab.—Florissant. U. S. Geol. Expl. Dr. *F. V. Hayden*.

ALNUS, Tourn.

"U. S. Geol. Rep.," vii. p. 139.

Alnus Kefersteinii, Goepp.

Ibid., p. 140, pl. xviii. figs. 6–8; lxiv. fig. 11.

Alnus inæquilateralis, Lesqx.

Ibid., p. 141, pl. lxii, figs. 1–4.

Alnus cordata, sp. nov.

Leaf cordate at base, pyramidal and acuminate, doubly serrate on the borders, long-petioled; primary nerves thick, the lateral opposite, parallel, 8 pairs, at acute angles of divergence, curving in passing to the borders, craspedodrome.

The leaf is 6 centimeters long, has a thick petiole 3 centimeters long, is largest near the cordate base (3 centimeters), and hence tapering to an acute point and dentate all around. The leaf resembles *Alnus diluviana*, Ung., "Iconogr.," pl. xvi, fig. 16, but is more acutely tapering to the point, and the lateral nerves, at a more acute angle of divergence, are more curved.

Hab.—Florissant. Lacoe's Cabinet, No. 83.

Flowers of *Alnus*, pl. xxxix, fig. 3, are also found at Florissant, but are not identifiable in species.

CUPULIFERÆ.

OSTRYA, Michx.

"U. S. Geol. Rep.," vii. p. 142.

Ostrya betuloides, sp. nov.

Leaves small, broadly ovate, acute, rounded to the equilateral base; borders dentate; lateral nerves close, at a broad angle of divergence.

The leaf is of the same size and shape as that of *Ostrya Atlantidis*, Sap., "Ét.," ii, 2, p. 254, pl. vi, fig. 4, differing in the simple teeth of the borders, which give to the leaf the appearance of a *Betula;* but there is with the same specimen a fragment of an involucre of *Ostrya*, similar in size to that of Sap., fig. 11, *l. c.*, and still more to *Ostrya tenerrima*. Sap., "Ét.," i, 2, p. 49, pl. v, fig. 6, differing only from the last by its larger size (2 centimeters long). Possibly this involucre is referable to the same species as the leaf. It is the only one seen, as yet, from this formation.

Hab.—Florissant. Lacoe's Cabinet, Nos. 26 and 29.

CARPINUS, Linn.

"U. S. Geol. Rep.," vii, p. 142.

Carpinus grandis, Ung.

Ibid., p. 143, pl. xix, fig. 9; lxiv, figs. 8–10.

Carpinus attenuata, sp. nov.

Plate XXVII, Fig. 10.

Leaf large, narrowed downward from the middle and upward to an acuminate point, slightly unequilateral at base; borders doubly dentate; lateral nerves oblique, straight, or slightly curved in passing up to the borders, branching near the borders, entering the primary teeth by their ends and the intermediate ones by their branches.

This leaf, 11 centimeters long, 5½ centimeters broad in the middle, its widest part, is equally narrowed upward and downward, with borders cut by large teeth entered by the secondary nerves, and generally two smaller ones intermediate or on the lower side of the primary teeth. The leaf appears to have been somewhat unequal at the base, but the broader side is lacerated; the veins are, however, equally oblique at the base and not more open on one side. The leaf closely resembles *Carpinus alnifolia*, Goepp., "Schoss. Fl.," p. 19, pl. iv, fig. 11, merely differing by the border teeth being a little larger, and by the more distinctly narrowed and elongated base. Schimper unites this last species to *C. ostryoides* of Goepp., *l. c.*, figs. 7–10. Fig. 7 represents a much smaller leaf, but it is narrowed to the base nearly in the same degree as in that of Florissant.

Hab.—Florissant. Princeton Museum. No. 258.

Carpinus fraterna, sp. nov.

Plate XXVII, Figs. 12–14.

Leaves small, lanceolate, rounded to the short petiole; borders minutely, sharply, doubly serrate; lateral nerves close, numerous, oblique and straight to the borders, branching near the borders.

The species is of the same type as *Carpinus Americana*, Linn., some of its varieties having leaves as small and of the same pattern. They are generally more coarsely or distinctly serrate than in the fossil species; the leaves are also generally larger.

Hab.—Florissant. U. S. Geol. Expl. Dr. *F. V. Hayden.*

FAGUS, Tournf.

"U. S. Geol. Rep.," vii, p. 115.

Fagus Feroniæ, Ung.

Ibid., p. 116, pl. xix. figs. 1–3.

QUERCUS, Linn.

Ibid., p. 117.

§ 1. Leaves dentate.

Quercus Haidingeri, Ett.

Ibid., p. 156, pl. xx, figs. 9, 10.

Quercus Mediterranea, Ung.

Plate XXVIII, Fig. 9.

Ung.. "Chlor. Protog.," p. 114, pl. xxxii, figs. 5–9; "Iconogr.," pl. xviii, figs. 1–6; Heer, "Fl. Tert. Helv.," ii, p. 52, pl. lxxiii, figs. 13, 15, 17, 18; Ung.. "Foss. Fl. v. Kumi," p. 25, pl. vi, figs. 1–22; Gaud., "Contr.," ii, p. 46, pl. iv, figs. 16–19.

Leaves coriaceous, obovate, abruptly acuminate, narrowed toward the base and abruptly rounded to it, deeply dentate; secondary nerves simple, craspedodrome, about 9 pairs; nervilles strong, at right angles to the secondary nerves, simple or more generally anastomosing in the middle.

Except that the teeth of the borders are slightly more acute and turned upward in the European species, I see no difference sufficiently marked to authorize a separation of this leaf into a new species. The leaf. fig. 3 of Ung., loc. cit., is like a counterpart of our fig. 9. and in other leaves figured by different authors the teeth of the borders are not sharply acute, but sometimes obtuse and nearly effaced. It is the case in Ung.. "Chlor.," pl. xxxii, fig. 5; in Heer, "Fl. Tert. Helv.," pl. lxxvi. figs. 13–15. The nervilles are distinctly seen in figs. 3–4 given of this species in Ung.. "Fl. v. Kumi," pl. vi. where twenty leaves of this species are represented. All these, however, have the border teeth more acute and proportionally smaller than in fig. 9 of our plate.

Hab.—Florissant.　U. S. Geol. Expl. Dr. F. V. Hayden.

Quercus serra, Ung.

"Chloris Protog.," p. 109, pl. xxx, figs. 5–7.

Leaves petioled, subcoriaceous, elliptical, pointed or obtuse, serrate-dentate on the borders; teeth equal, with callous points.

A single leaf. 4 centimeters long without the petiole. 2½ centimeters broad, remarkably similar to fig. 7 of Ung., loc. cit., oval or obtusely ovate,

with a short thick petiole. The lateral nerves are much curved in passing
to the borders, close, craspedodrome.

Hab.—Florissant. Lacoe's Collection, No. 64.

Quercus Drymeja, Ung.

Plate XXVIII, Fig. 12.

"U. S. Geol. Rep.," p. 157, pl. xix, fig. 14.

Among the numerous figures given of this species this leaf is especially comparable to Ung., "Chlor. Prot.," pl. xxxii, fig. 1, and to "Fl. of
Sotzka," pl. ix, fig. 1. The lateral veins are mostly craspedodrome, the
lower pairs entering the teeth by an anastomosing veinlet. The species
is very common in the Miocene of Europe. The reference of the fragment of leaf described, vol. vii, *loc. cit.*, is not certain.

Hab.—Randolph Co., Wyoming. U. S. Geol. Expl. Dr. *F. V. Hayden*

Quercus Osbornii, sp. nov.

Plate XXXVIII, Fig. 17.

Leaf small, obovate, abruptly long-acuminate, dentate from under the acumen
to the middle; medial nerve thin; secondary nerves oblique, alternate, parallel, camptodrome.

This fine leaf, about 7 centimeters long, is gradually narrowed from
above the middle to the base (broken), rounded in the upper part, there
cut by three or four large teeth, and then abruptly long-acuminate. The
lateral nerves diverging 30° to 40°, curve in passing up to the borders,
which they follow in festoons, entering the teeth by anastomosing branchlets. I do not find any other species comparable to this but *Quercus
Tephrodes*, Ung., as described in "Sieber, Nord-Böhm. Braun-Kohl.," pl.
iii. fig. 17. *Quercus hexagona*, Lesqx., "U. S. Geol. Rep.," vi, pl. v, fig. 8, is
also of the same type.

Hab.—Florissant. Princeton Collection, No. 684.

Quercus pyrifolia, sp. nov.

Plate XXVIII, Fig. 14.

Leaves rather thin, oval, short-acuminate, rounded in narrowing to a long petiole;
borders irregularly obscurely serrate; secondary nerves curving in passing to the borders, camptodrome, crossed by nervilles at right angles.

The petiole of the leaf is 1½ centimeters long, and the leaf without it is 5 centimeters long and nearly 3 centimeters broad in the middle. It is broken at the apex, but appears as tapering to a short acumen. The lateral nerves, 5 or 6 pairs, at an angle of 40°, are thin, flexuous, camptodrome, following the borders and joined to some of the teeth by anastomosing veinlets; nervilles flexuous or transversely curved.

Species related to *Quercus larguensis*, Sap., "Ét." iii, 1, p. 67, pl. 5, fig. 1, which has the same form, the borders irregularly cut-dentate.

Hab.—Florissant. Princeton Museum, No. 797.

Quercus castaneopsis, sp. nov.

Plate XXVIII, Fig. 10.

Leaves large, lanceolate, gradually acuminate, regularly distantly dentate; lateral nerves parallel, at an open angle of divergence, the lower joining the medial nerves at right angles, all camptodrome, curving in passing to the borders, following them and entering the short teeth by oblique nervilles; areolation of minute polygonal meshes.

This leaf may represent a *Castaneopsis*. I do not know of any fossil species to which it may be compared.

Hab.—Randolph Co., Wyoming. U. S. Geol. Expl. Dr. *F. V. Hayden.*

§ 2. Leaves entire.

Quercus elæna, Ung.

Plate XXVIII, Figs. 11, 13.

Ung., "Chlor. Protog.," p. 112, pl. xxxi, fig. 4; Heer, "Fl. Tert. Helv.," ii, p. 47, pl. lxxiv, figs. 11–14; lxxv, fig. 1; iii, p. 178, pl. cli, figs. 1–3; Sap., "Ét.," ii, p. 85, pl. iii, fig. 11; iii, p. 65, pl. ii, figs. 5–9; v, fig. 2.

Leaves coriaceous, short-petioled, oblong-lanceolate; borders entire, revolute or reflexed; lateral nerves camptodrome.

The leaves vary from 5 to 7 centimeters long and from 1 to 1½ centimeter broad. Those figured here especially resemble the figures in Sap., *loc. cit.*, pl. ii, figs. 5–10.

Hab.—Florissant. U. S. Geol. Expl. Dr. *F. V. Hayden.*

Quercus neriifolia, Al. Br.

Plate XXXI, Fig. 12.

"U. S. Geol. Rep.," vii, p. 150, pl. xix, figs. 4, 5.

I refer with doubt to this species a subcoriaceous polished leaf 10 centimeters long, 22 millimeters broad in the middle, whose borders are not entire but distantly dentate, and the base slightly decurrent to a thick short petiole. In the European species the leaves are mostly entire, but sometimes also denticulate in the upper part, and the base of the leaf is not as decurrent, while the petiole, generally thick, is a little longer. The nervation is as represented in Heer, "Fl. Tert. Helv.," ii, pl. lxxiv, fig. 4.

Hab.—Randolph Co., Wyoming.　U. S. Geol. Expl. Dr. *F. V. Hayden.*

CASTANEA, Linn.

Castanea intermedia, Lesqx.

"U. S. Geol. Rep.," vii. p. 164, pl. xxi, fig. 7.

SALICINEÆ.

SALIX, Linn.

Salix amygdalæfolia, sp. nov.

Plate XXXI, Figs. 1, 2.

"U. S. Geol. Rep.," p. 165.

Leaves narrowly lanceolate, tapering to a blunt acumen, rounded in narrowing to the petiole, serrulate; lateral nerves at an acute angle of divergence.

The leaves, 6 to 7 centimeters long, 12 to 15 millimeters broad, with a slender petiole 2 centimeters long, may seem to represent a variety of *S. varians,* Goepp., so common in the European Miocene. But they are generally much smaller, more narrowly lanceolate; the secondary nerves, especially the basilar ones, at a more acute angle of divergence; the borders more distinctly serrate-crenate. The form of the leaves is the same as in *S. larateri,* Al. Br., but the leaves of this last species are much longer.

Hab.—Florissant. Seen in the different collections from that locality.

Salix Libbeyi, sp. nov.

Plate XXXI, Fig. 3.

Leaves large, thick, oblong, enlarged upward, rapidly narrowed to the point, tapering to the base, very entire.

The nearest relation of this species is *S. abbreviata,* Goepp., "Schoss. Fl.," p. 25, pl. xvii, figs. 4–11, especially like fig. 7; but the American leaf is twice as large, 8 centimeters long, 2½ broad in the upper part, narrowed to the base, which is not rounded, and more enlarged upward.

Hab.—Florissant.　Princeton Museum, No. 780.

Salix media, Heer.

"U. S. Geol. Rep.," vii. p. 168, pl. xxii, fig. 3.

Salix angusta, Al. Br.

Ibid., p. 168, pl. xxii, figs. 4, 5.

Salix elongata, O. Web.

Ibid., p. 169, pl. xxii, figs. 6, 7.

POPULUS, Linn.

Ibid., vii. p. 169.

Populus Heerii, Sap.

Plate XXX, Figs. 1-8; XXXI, Fig. 11.

Sap., "Ét.," i, p. 87, pl. vii, fig. 3.

Leaves long-petioled, ovate, long-lanceolate, acuminate, obtusely serrate; primary nerves thick; lower secondary nerves at a more acute angle of divergence and ascending higher along the borders, the others curving in passing to the borders and reticulate in following them.

The leaves are extremely variable in size, some, as shown in fig. 5, being 20 to 30 centimeters long and 10 to 12 centimeters broad below the middle; others, as in fig. 2, scarcely 5 centimeters long and 2 broad; others still, as in fig. 11 of pl. xxxi, being narrow comparatively to their length, 10 centimeters long, 2 centimeters broad, thus resembling leaves of willows. That all these leaves represent the same species is evident enough. Besides the essential characters in common, they have the same somewhat thick consistence, and are all colored reddish-yellow even upon shales where all the fragments of other plants are colored black.

Saporta, who has described a fruit of *Populus* found upon the same slate as his leaf, compares it to that of *P. Euphratica*, Oliv., and the leaves to *P. laurifolia*, Ledeb. We have still living in the Rocky Mountains of Colorado and Utah a species, *P. angustifolia*, James, considered by some authors as a variety of *P. balsamifera*, Linn., which represents the fossil species in the different forms and size of its leaves. Those of the living species vary from 5 to 24 centimeters long and 2 to 10 centimeters broad, being either attenuated or broadly cordate at base, according to their width.

Hab.—Florissant. Found in all the collections.

Populus balsamoides? Goepp., var. latifolia.

Plate XXXI, Fig. 4.

Goepp., "Fl. v. Schoss.," p. 23, pl. xv, figs. 5, 6; Heer, "Fl. Tert. Helv.," ii, p. 18, pl. lix; lx, figs 1-3.

Leaf very large, apparently broader than long, cordate-ovate; borders undulate, crenate; primary nerves thick; lateral nerves thin, much curved to and along the borders: the lower pairs much branched, the other simple.

This leaf, about 12 centimeters long and 14 broad toward the base, seems to represent a different species from those figured under this name by European authors. It is broader than long, while the leaves of *P. balsamoides* are, according to Heer, always longer than broad; it is deeply cordate at base, and the lateral veins, without any basilar veinlets, are comparatively very thin, much curved and all alike; the borders are merely crenulate, even obscurely so, while they are more or less deeply serrate in the normal form of *P. balsamoides*. Fig. 7, pl. lix, of Heer, *l. c.*, represents, however, a leaf with borders obscurely dentate and nearly as large as that of fig 4, cordate at base; and fig. 1 of pl. lx of Heer shows the lateral nerves of the same character as they are in the American leaf. There is between the fossil leaves a difference as marked as between those of the living *Populus balsamifera*, Linn., and *P. candicans*, Ait. This last, though with broader and more or less heart-shaped leaves, is considered a mere local variety of the first.

Hab.—Florissant.   U. S. Geol. Expl. Dr. *F. V. Hayden.*

Populus Zaddachi. Heer.

Plate XXXI, Fig. 8.

"U. S. Geol. Rep.," vii, p. 176, pl. xxii, fig. 13.

The figured leaf is one of the smallest of this species, and besides differs from the normal form in some points. The secondary nerves descend a little lower; the border teeth, though obtuse and turned upward, have not at the apex the small glands which are generally seen in the small leaves of this species. As these glands may have been destroyed by maceration, as is often the case, and as this species is very common in the North American Tertiary, I consider this leaf as a mere variety.

Hab.—Florissant.   U. S. Geol. Expl. Dr. *F. V. Hayden.*

Populus oxyphylla, Sap.

Plate XXXVIII, Figs. 9-11

Sap., "Ét.," iii, 1, p. 73, pl. vii, fig. 1.

Leaves of small size, long petiolate, deltoid, short-acuminate, rounded to the base, denticulate; secondary nerves variable in distance, the lower longer, branching outside.

The leaves vary from 2½ to 4 centimeters long and from 1½ to 2½ centimeters broad below the middle, from which part they taper upward to a point or short acumen; the petiole is 2 to 3 centimeters long. The author describes and figures the lateral nerves as flexuous, a character which is not seen on the leaves which I refer to this species. The nerves are, however, camptodrome, the teeth being entered, as seen in fig. 11, the best preserved leaf, by short veinlets anastomosing to the curves of the lateral nerves. In this leaf also the nervilles and their mode of ramification in forming large primary irregularly hexagonal meshes are of the same type as in the figure of Saporta.

Hab.—Florissant. U. S. Geol. Expl. Dr. *F. V. Hayden.* One specimen, No. 54, not figured here, is in the collection of Mr. Lacoe.

Populus Richardsoni, Heer.

"U. S. Geol. Rep.," vii, p. 177, pl. xxii, figs. 10-12.

Populus arctica, Heer.

Ibid., p. 178, pl. xxiii, figs. 1-6.

BALSAMIFLUÆ.

LIQUIDAMBAR, Linn.

Ibid., vii, p. 186.

Liquidambar Europæum, Al. Br.

Plate XXXII, Fig. 1.

Al. Braun, "Buckl. Geol.," p. 112; Ung., "Chlor. Protog.," p. 120, pl. xxx, figs. 1-5; Goepp., "Tert. Fl. v. Schoss.," p. 22, pl. xii, figs. 6,7; Heer, "Fl. Tert. Helv.," ii, p. 6, pl. li, lii, figs. 1-8; Ludw., "Palæontog.," viii, p. 89, pl. xxv, figs. 1-4; Gaud., "Contrib.," iv, p. 19, pl. iv, figs. 5-7.

Leaves long-petioled, palmately 3 to 5-lobed; lobes more or less distinctly glandulose, serrulate, lanceolate-acuminate.

In the leaf figured as referable to this species the borders appear nearly entire or merely undulate-crenate; but it is the only difference from the normal form which is very common in the Miocene of Europe.

The leaves preserved flattened on some of the thin sandy shales of Florissant very often have the borders erased and the small teeth therefore often
destroyed. The medial lobe of the figure has the teeth quite as distinct
as in some of the figures of European authors, still more so than in fig. 5
of Gaudin, *l. c.*

Hab.—Randolph Co., Wyoming.   U. S. Geol. Expl.   Dr. *F. V. Hayden.*

URTICINEÆ.

ULMACEÆ.

ULMUS, Linn.

"U. S. Geol. Rep.," vii, p. 187.

Ulmus tenuinervis, Lesqx.

Ibid., p. 188, pl. xxvi, figs. 1, 3.

Ulmus Hilliæ, sp. nov.

Plate XXVIII, Figs. 1, 3.

Leaves narrow, lanceolate-acuminate, very unequilateral at base, simply or
doubly-serrate; lateral veins curved in passing to the borders, craspedodrome.

The leaves are small, 5 to 9 centimeters long, 1½ to 2½ centimeters
broad, short-petioled, thickish; the base is narrowed on one side in
rounding to the petiole, straight on the other; the teeth of the borders are
large, slightly turned up, not very sharp; the areolation is quite distinct
in small irregularly quadrangular meshes, formed by subdivisions of nervilles mostly at right angles.

Hab.—Florissant. Mrs. *Hill,* who has widely collected and distributed the specimens of fossil plants of that locality.

Ulmus Brownellii, sp. nov.

Plate XXVIII, Figs. 2, 4.

Leaves narrow, oblong-lanceolate, unequal at base, simply obtusely dentate;
lateral nerves simple, parallel, the lower open; nervilles irregularly branching and
anastomosing; areolation polygonal, loose.

This species resembles the preceding, differing by the simple teeth
and nerves; the areoles, much larger, formed by irregularly divided
nervilles.

Hab.—Florissant. U. S. Geol. Expl.; White River. *W. A. Brownell.*

Ulmus Braunii, Heer.

Plate XXVII, Figs. 1–1.e.

Heer, "Fl. Tert. Helv.," ii. p. 59, pl. lxxix. figs. 14–21 ; iii. p. 181. pl. cli, fig. 31 ; Gaud., "Contrib.," ii, p. 47, pl. iii, figs. 3–9 ; Ludw., "Palæontog.," vii. p. 105, pl. xxxviii. figs. 5–8 ; Ett., "Fl. v. Bil.," p. 61, pl. xviii, figs. 23–26

Leaves short-petioled, very unequilateral, round or cordate at base, elliptical or ovate-lanceolate, acute or acuminate, doubly or simply coarsely dentate; teeth conical, turned up; lateral veins open, at right angles toward the base, 12–18 pairs; fruit peti-olate, broadly-winged; wings lateral.

This species is very variable in the form of the leaves and the more or less acute teeth of the borders. The leaves, 4½ to 12 centimeters long, 2½ to 4½ centimeters broad, are comparatively broader and shorter and more unequilateral and difform than those of the preceding species. It is very common in the European Miocene and is also abundantly found at Florissant, where the fruits also are not rare. But these fruits, always found ripe, do not agree with the figures given by Heer, *loc. cit.*. pl. cli, fig. 31 ; they are rather like those of *U. Brownii,* or *U. longifolia.* Ung., as figured in "Bil. Fl.," pl. xviii, figs. 4, 5, 8. The specific relation of the seeds of *Ulmus* described by European authors is hypothetical, as well as that of those I have figured.

Hab.—Florissant. Not rare; especially in Princeton Collection.

PLANERA, Gmel.

"U. S. Geol. Rep.," vii, p 189.

Planera longifolia, Lesqx.

Plate XXIX, Figs. 1–13; XLIV, Fig. 10.

Lesqx., "U. S. Geol. Rep.," vii. p. 189, pl. xxvii, figs. 4–6.

Planera longifolia, var. myricæfolia.

Plate XXIX, Figs. 15–27.

From a comparison made in the examination of more than two thousand specimens, representing not merely the leaves figured but a large number of intermediate forms, I have been forced to admit that they all belong to the same species, and that though some of them are closely allied to the European *Planera Ungeri,* they constitute a different species. First examining the relation of all the leaves from No. 1, the normal type, to

C F 11

No. 13, all have simple, more or less acute, more or less distant teeth;
and the lateral veins all simple, straight, craspedodrome, vary in nothing
but in their more or less acute angle of divergence according to the width
of the leaves; the petiole is equally variable, from 5 to 10 millimeters
long, and the leaves are sometimes nearly sessile, as in fig. 7.   One of the
leaves of fig. 1 has also the petiole very short.   Comparing the different
forms of figs. 14–27 we see the same essential characters preserved—that
is, lateral veins straight, craspedodrome, at a more or less acute angle of
divergence relatively to the width of the leaves, the teeth either sharply
acute, even acuminate, or merely pointed, even obscurely so, as in figs.
25, 27.   The petiole is generally of the same length, but some of the leaves
(figs. 21, 26, 27) are narrowed to the base and nearly without petiole.
If I add that all these leaves have the same consistence and black color
upon the shale, that both forms are often found upon the same specimens,
that it is often scarcely possible to say that a leaf is referable to the normal type or to the variety, it will be understood why I am unable to consider these leaves as representing different species or referable to two
genera, though, comparing the extreme forms (figs. 1, 5, 6, to figs. 21, 24,
27), this separation seems indeed natural.

As for the identity of this species with *P. Ungeri*, it is disproved by the
comparatively large and narrower leaves, the veins, exactly straight from
the medial nerves to the point of the teeth, never curved, and the fruits
which, as seen in comparing fig. 12 with fig. 1, pl. lxxx of Heer, "Fl. Tert.
Helv.," are nearly twice as large in the American species.   The difference
in the characters of the leaves may be easily seen in comparing the figures
of pl. xxix with that of *P. Ungeri*, quoted below.

Hab.—Florissant.   Most abundant.

Planera Ungeri, Ett.

"U. S. Geol. Rep.," vii, p. 190, pl. xxvii, fig. 7.

CELTIDEÆ.

CELTIS. Tourn

"U. S. Geol. Rep.," vii, p. 191.

Celtis McCoshii, sp. nov.

Plate XXXVIII, Figs. 7, 8.

Leaves long-petioled, narrowly ovate, lanceolate-acuminate, more or less unequilateral at base; lower lateral nerves at a more acute angle of divergence, ascending higher across the borders, curved like the upper (4 to 6 pairs), all camptodrome, attached to the borders by anastomosing veinlets.

The leaves, 5 to 6½ centimeters long, 2 to 2½ centimeters broad below the middle, where they are widest, are not very but distinctly unequilateral at the rounded base, at least in fig. 7. By the form of the leaves the species is closely allied to *Celtis primigenia*, Sap., "Ét.," ii. 2, p. 263, pl. vi, fig. 7. The nervation and the denticulation of the leaves are of the same character. The leaves are also remarkably similar to those of *C. occidentalis*, Linn., var. *Texana*, a form whose leaves, nearly equilateral at base, are minutely serrate. The Texas leaves are subcordate at base or round, as in fig. 8.

Hab.—Florissant and Randolph Co., Wyoming. Princeton Collection. No. 794, U. S. Geol. Expl. Dr. *F. V. Hayden.*

MOREÆ.

FICUS, Tourn.

" U. S. Geol. Rep.," vii, p. 191.

Ficus lanceolata, Heer.

Ibid., p. 192, pl. xxviii. figs. 1, 5.

Ficus Jynx, Ung.

Ibid., p. 193, pl. xxviii, fig. 6.

Ficus multinervis, Heer.

Ibid., p. 191, pl. xxvii, figs. 7, 8.

Ficus arenacea, Lesqx.

Ibid., p. 195, pl. xxix, figs. 1–5.

Ficus Ungeri, Lesqx.

Plate XLIV, Figs. 1–3.

Ibid., p. 195, pl. xxx, fig. 3.

This species is finely represented by the three figures of our plate. They show not merely the variable size of the leaves, but their true shape and the short petiole abruptly thickened at base. The leaves, are oblong or lingulate, rounded at the base and apparently at the apex also; they vary in size from 10 to 20 centimeters long and from 3½ to 6½ centimeters broad in the middle. Fig. 2 may represent a different species not merely on account of the different size, but from the presence of tertiary thinner and shorter veins intermediate to the secondary nerves.

Hab.—Alkali Station, Wyoming. Professor *Scudder:* Green River Station. U. S. Geol. Expl. Dr. *F. V. Hayden.*

Ficus Wyomingiana, Lesqx.

"U. S. Geol. Rep." vii. p. 205. pl. xxxiv, fig. 3.

Ficus tenuinervis, sp. nov.

Plate XLIV, Fig. 4.

Leaf oblong or lanceolate, tripalmately nerved, rounded at base, entire.

A mere fragment, showing the lower part of a leaf whose lower lateral nerves are strongly branched downward and all (nerves and branches) camptodrome. The medial nerve is inflated at base. The fragment represents a *Ficus*, but the specific characters are not discernible.

Hab.—Alkali Station. Professor *Scudder.*

Ficus alkalina, sp. nov.

Plate XLIV, Figs. 7–9.

Leaves thin, variable in size, obovate or ovate-lanceolate, acuminate, obtusely serrulate, palmately trinerved; secondary nerves distinct, all camptodrome, alternate and parallel; nervilles oblique, simple or forking in the middle.

The leaves are fragmentary, variable in length from 6 to 10 centimeters, and proportionally broad. The nervation is that of a *Ficus:* the lower primary lateral nerves are thin, flexuous, ascending at a more acute angle of divergence. The upper are parallel, camptodrome, attached to the teeth by small anastomosing nervilles.

Hab.—Alkali Station. Professor *Scudder.*

SANTALEÆ.

SANTALUM, Linn.

Santalum Americanum, sp. nov.

Plate XXXII. Fig. 7.

Leaves thick, narrowly elliptical or oblong, very short-petioled, blunt at the apex: nervation obsolete.

The basilar border of the leaf is decurrent along the petiole, which is scarcely 2 millimeters long for a leaf 4 centimeters long, 1 centimeter broad in the middle. The affinity of this leaf is with the living *Santalum lanceolatum*, Brown. From the fossil species published, it differs in the very short petiole and the blunt apex of the leaves.

Hab.—Florissant. No. 638 of the collection of the Princeton Museum.

LAURINEÆ.

CINNAMOMUM, Burn.

Cinnamomum Scheuchzeri, Heer.

Plate XXXVIII, Fig. 6.

"U. S. Geol. Rep.," vii, p. 220, pl. xxxvii, fig. 8.

The leaf from Florissant more distinctly represents this species than that ("Rep." vii) from Montana. There is still a small difference from the European form in the position of the lateral nerves descending lower, nearly to the top of the petiole, and the basilar borders more distinctly decurrent. These deviations from the normal character are, however, somewhat indicated in a few of the numerous figures given by Heer of this species.

Hab.—Florissant. U. S. Geol. Expl. Dr. *F. V. Hayden.*

PROTEACEÆ.

BANKSITES. Sap.

Banksites lineatus, sp. nov.

Plate XXXII, Fig. 21.

Seeds obliquely oval, winged; wings oblong, obtuse, larger on one side, distinctly striate lengthwise by 5 or 6 parallel black lines converging at the apex.

The seeds resemble those described as *Banksia Radobojensis*, Ung., "Syllog.," iii, p. 75, pl. xxiv, figs. 16, 17.

Hab.—Florissant; not rare, but as yet no leaves referable to this genus have been found there.

LOMATIA, R. Br.

Leaves coriaceous, pinnately laciniate or acutely lobed; divisions oblique, lanceolate, acute or acuminate, nerved in the middle, decurrent along the medial nerve or connected by a narrow wing at the basilar margin.

This definition merely relates to the peculiar leaves described below, whose relationship is marked only with leaves of some species of *Lomatia*. Their texture is thick. The surface is always covered by a coaly layer, obliterating the nervation.

Lomatia hakeæfolia, sp. nov.

Plate XXXII, Fig. 19.

Leaf obliquely truncate at base, lanceolate, acuminate, irregularly deeply dentate.

This form differs from the following by the segments, or lobes, being shorter and directed to the outside at right angles to the primary nerve; these acute short lobes or teeth, four on each side, are opposite and separated by broad shallow sinuses; no trace of secondary nerves is discernible.

Hab.—Florissant; rare. U. S. Geol. Expl. Dr. *F. V. Hayden.*

Lomatia spinosa, sp. nov.

Plate XLIII, Fig. 1.

Leaves narrowly lanceolate, long-acuminate, broadly alternately acutely dentate-lobed; divisions gradually shorter upward, the terminal long-acuminate.

Related to the preceding species but differing by the laciniæ being longer, turned upward, decurrent. The primary nerve is scarcely visible.

Hab.—Florissant; rare. U. S. Geol. Expl. Dr. *F. V. Hayden.*

Lomatia terminalis, sp. nov.

Plate XLIII, Figs. 2–7.

Leaves linear-lanceolate, acuminate, deeply lobate; lobes oblique, lanceolate, acute, decurrent along the primary thin nerve; lateral nerves generally distinct.

Hab.—With the preceding; not rare. U. S. Geol. Expl. Dr. *F. V. Hayden.*

Lomatia tripartita, sp. nov.

Plate XLIII, Figs. 8–10.

Leaves palmately trilobate, narrowly cuneate to the base; lobes obliquely diverging, oblong, obtuse or obtusely pointed, entire or dentate-lobed on one side; primary nerves more or less distinct.

The three fragments representing this species may be mere forms of the preceding.

Hab.—Florissant; rare. U. S. Geol. Expl. Dr. *F. V. Hayden.*

Lomatia acutiloba, sp. nov.

Plate XLIII, Figs. 11–16, 20.

Leaves long, linear-lanceolate, alternately pinnately lobed; lobes lanceolate or linear-lanceolate, acute, oblique, decurrent, gradually shorter upward, distinctly curved backward.

The divisions of the leaves, their shape and mode of decurring to a primary axis, are of the same type as in *Lomatia (Todea) Saportana* of the "Cretaceous Flora" ("U. S. Geol. Rep."), vi, pl. xxix, figs. 1–4.

Hab.—Florissant. Common, and seen in all the collections.

Lomatia abbreviata, sp. nov.

Plate XLIII. Fig. 17.

Leaves linear or narrowly lanceolate; lobes oblique, short, oblong, not decurrent, cuneate at base, inclined upward, obtusely pointed; nerves obsolete.

This fragment appears related to fig. 10.

Hab.—Florissant; very rare. Collection of the Princeton Museum.

Lomatia interrupta, sp. nov.

Plate XLIII, Figs. 18, 19.

Leaves linear-oblong, larger in the middle, either lobes bi-form; larger, ovate, entire or obtusely dentate, or smaller intermediate to the larger ones, merely oval-obtuse, like short teeth.

This peculiar form has the lobes of the top and the base of the leaves simple, open, obtuse; in the middle the lobes become larger, obovate, obtusely irregularly dentate, opposite, and near their base the wing of the leaves is expanded into intermediate very small entire obtuse teeth. The large lobes, when entire, have only the medial nerve distinct; in the dentate ones the medial nerve is dichotomous, the branches passing up to the teeth, one or two on each side.

Hab.—Florissant; very rare. Princeton Collection, Nos. 842, 843.

Lomatia microphylla, Lesqx.

"U. S. Geol. Rep.," vii, p. 211, pl. lxv, figs. 14, 15.

PIMELEÆ.

PIMELEA. Banks.

Pimelea delicatula, sp. nov.

Plate XXXIII, Figs. 15, 16.

Leaves membranaceous, nearly sessile, spatulate, short-pointed or apiculate; secondary nerves emerging at an acute angle of divergence, branching on the lower part, variable in distance, separated by intermediate short veinlets; nervation camptodrome.

The leaves vary from 3 to 5½ centimeters long and from 8 to 13 millimeters broad in the upper part, near the apex, where they curve upward in narrowing to a short point, and from which part they are gradually narrowed downward to the very short petiole.

The species is closely allied to *P. Œningensis*, Heer, "Fl. Tert. Helv.," ii, p. 93, pl. xcvii, figs. 2–10, which has smaller leaves less gradually narrowed downward and no petiole.

Hab.—Florissant.   U. S. Geol. Expl. Dr. *F. V. Hayden.*

OLEACEÆ.

OLEA. Linn.

Of the numerous living species of this genus, one only, *Olea Americana*, inhabits the North American Continent; three species are European: the others are found in Tropical Asia and South Africa: Japan has one species.

The leaves of *Olea* are opposite, petioled, coriaceous, persisting, oblong-oval, obovate or lanceolate, very entire; the nervation pinnate, and the flowers fasciculate in the axils of the leaves.

Olea præmissa, sp. nov.

Plate XXXIII, Fig. 1.

Leaves coriaceous, lanceolate, larger below the middle, narrowed to a very short petiole; flowers in simple or rarely compound racemes.

The leaves average 5 centimeters in length and 1 centimeter in width below the middle, from which they are gradually tapering upward to a blunt point. The flowers are short-petioled, either single or in short slightly compound racemes. This character essentially separates this species from *Olea Americana*, its nearest relative, from which it differs by smaller leaves and larger flowers. No trace of secondary veins is discernible on those leaves.

Nine fossil species of *Olea* are described by authors from the Miocene of Europe, none of which have a marked relation to this.

Hab.—Florissant. Princeton Collection. No. 641.

FRAXINUS, Tourn.

"U. S. Geol. Rep.," vii. p. 228.

Fraxinus prædicta, Heer.

Ibid., p. 229, pl. xl, fig. 3.

Fraxinus Heerii, sp. nov.

Plate XXXIII, Figs. 5, 6.

Leaflets more or less unequilateral, rounded or narrowed to the short petiole, and equally so from the middle to the acuminate blunt apex; borders undulate; lower secondary nerves at a more acute angle of divergence, all unequally distant, curving and reticulate at a distance from the borders; nervilles flexuous, at right angles to the medial nerve.

The leaflets, 5 to 7 centimeters long, 1½ to 2 centimeters broad, are, evidently, part of a compound leaf, as seen from the lower lateral leaflet, which is nearly sessile and very unequilateral, and the upper a terminal one, equilateral, larger and petioled. The lateral nerves are thin, arched toward the medial nerve at a distance from the borders, as in *Fraxinus prædicta*, Heer, "Fl. Tert. Helv.," pl. civ, figs. 12, 13, to which this species is closely related; indeed, it merely differs by the basilar nerves being at a more acute angle of divergence, and those above with curves more distant from the margins which are merely undulate. No fruiting part has been found.

Hab.—Florissant. U. S. Geol. Expl. Dr. *F. V. Hayden.*

Fraxinus mespilifolia, sp. nov.

Plate XXXIII, Figs. 7–12.

Leaflets more or less unequilateral, ovate-lanceolate, obtusely acuminate, rounded to a short petiole, obtusely serrate; secondary nerves parallel, subequidistant, 8 or 9 pairs, much curved in passing to the borders and following them, connected with the teeth by short anastomosing veinlets; nervilles oblique, very flexuous.

This species is as closely allied to *F. juglandina*, Sap., "Ét.," iii. p. 89, pl. ix. figs. 13–16, as is the preceding to *F. prædicta*, Heer. The leaflets are broader, less unequal than in *F. Heerii*, rounded or narrowed on one side to a short petiole; the camptodrome veins follow close to the borders, not curving inside to the medial nerves, and the borders are always distinctly serrate. In *F. juglandina* the borders are sharply denticulate and the more open lateral veins do not ascend higher along the borders, as in the American species.

Hab.—Florissant.   U. S. Geol. Expl. Dr. *F. V. Hayden*.

Fraxinus abbreviata, sp. nov.

Plate XXVIII, Figs. 5, 6.

Leaves short, ovate, acute, round or truncate at base, short-petioled, denticulate; secondary nerves close, parallel, open, curved in passing to the borders, much branching outside.

These leaflets, subequilateral, 3 to 5 centimeters long, 2 to 3 centimeters broad, with borders equally cut in acute small teeth slightly turned upward, have the lateral nerves close, 10 pairs, at an angle of divergence of 60°, somewhat curved in traversing the areas, much divided near the borders, the branches entering the teeth directly or by anastomosing veinlets. The nervation is like that of *Fraxinus ulmifolia*, Sap., "Ét." iii. p. 91. pl. ix. figs. 17–19, differing essentially by shorter, comparatively broader, more equilateral leaflets, and less acute, more equal teeth. The relation of the species is very close.

Hab.—Florissant.   U. S. Geol. Expl. Dr. *F. V. Hayden*.   Seen also in Lacoe Cabinet, No. 26.

Fraxinus? myricæfolia, sp. nov.

Plate XXXIII, Figs. 13, 14

Leaflets small, sessile, subcoriaceous, narrowly lanceolate, distantly dentate; secondary nerves very oblique, mostly obsolete.

The relationship of this fragment of leaf is obscure. The lateral nerves are obsolete and the leaflets sessile. Though the leaflet, fig. 14, has the same thick texture, the nerves scarcely distinct, it seems different on account of its short petiole and the direction of the secondary nerves, which is at an acute angle of divergence, apparently toward the teeth as craspedodrome. It may be a leaf of *Myrica*.

Hab.—Florissant.   U. S. Geol. Expl. Dr. *F. V. Hayden*.

Fraxinus Ungeri, sp. nov.

Leaflet small, membranaceous, very entire, unequilateral, broadest below the middle, ovate-lanceolate, acuminate, narrowed to a short petiole.

There are three leaflets of the same kind remarkably similar in shape and size to *Fraxinus primigenia*, Ung., "Syllog.," i. p. 22. pl. viii. figs. 3–8. They are 4½ to 7 centimeters long, 1½ to 2½ centimeters broad below the middle, where they are much larger on one side than the other. The secondary nerves are parallel, open, curved in traversing the areas, branching near the borders, effaced in touching them. It may be the same species as that of Unger, but it is not possible to ascertain the degree of relationship, as in the leaflet representing the European species the secondary nerves are neither described nor distinctly figured.

Hab.—Florissant. Lacoe's Cabinet, No. 57.

Fraxinus Brownellii, Lesqx.

"U. S. Geol. Rep.," vii. p. 230.

Fraxinus Libbeyi, sp. nov.

Plate XXVII, Figs. 5–7, 9.

Leaves very variable in size, unequilateral, ovate-lanceolate, acuminate, rounded to a short petiole, irregularly serrate; secondary nerves parallel, close, 10 to 18 pairs according to size, branching near the borders, camptodrome, joined to the teeth by anastomosing veinlets.

The leaves vary from 3½ to 11 centimeters long, 1½ to 4 centimeters broad. They are very unequal at base, generally cut straight and obliquely on one side toward the petiole, enlarged and rounded on the other, deeply more or less irregularly serrate. Fig. 9 represents a long narrow leaf, broader in the middle, gradually narrowed upward and downward, rather oblong; the other leaves are broader toward the base and ovate; the secondary nerves are more or less divided near the borders, generally camptodrome, joined to the teeth by nervilles, a few of them entering the teeth; the nervilles are parallel, flexuous, simple or forking, or anastomosing at right angles in the middle; the areolation as seen in fig. 9 is formed of very small quadrate or round-quadrangular meshes.

Hab.—Florissant. Princeton Museum, Nos. 217, 245, 275, 281.

APOCYNEÆ.

APOCYNOPHYLLUM, Ung.

Leaves very entire, penninerve, coriaceous: medial nerves strong; secondary nerves very open or at right angles to the midrib, close together, camptodrome, sometimes separated by shorter intermediate thin veins.

Apocynophyllum Scudderi, sp. nov.

Plate XLV*, Figs. 1-5.

Leaves oblong-lanceolate, gradually narrowed upward to an acumen and downward to a short petiole; secondary veins nearly at right angles, numerous, camptodrome, and curving quite near and along the borders as if joined to a continuous lateral nerve ; intermediate tertiary nerves thinner, as long as the secondary ones; nervilles close, oblique.

The peculiar direction of the nerves, which in their curves follow the borders, appearing like a continuous marginal vein, is also a character of the leaves of some *Myrtaceæ*. The relationship of this species is, however, more marked, not only by the nervation but by size and form of the leaves with *Apocynophyllum Helveticum*, Heer, figured in "Bornst. Fl.," pl. iv. figs. 1-7. The curving of the veins close to the borders is distinctly seen (fig. 3) with the intermediate tertiary nerves, corresponding to fig. 4 of Heer. The form of the leaves and their size being also the same, possibly the American species is a mere variety.

Hab.—Alkali Station. Professor *Scudder*.

CONVOLVULACEÆ.

PORANA, Burm.

I have seen of this genus scariose calyxes, but, as yet, no leaves. These calyxes, 3- to 5-lobate, have the sepals generally of unequal length, free to the base, sometimes more or less connate. Two species only are described by authors with calyxes and leaves, six from scariose calyxes, all from the European Miocene.

Porana Speirii, sp. nov.

Plate XXVIII, Fig. 15.

Calyx scariose, somewhat thick, indistinctly five-lobate; lobes large, connate; nerves diverging from the central point to the borders, traversed at right angles by strong nervilles, forming equilateral meshes.

The lobes are marked only by their upper borders being connate to near the rounded apex, where they are more than 1½ centimeters broad and of the same length. This form is related to *Getonia membranosa*, Goepp., "Schoss. Fl.," p. 38, pl. xxv. fig. 12, whose sepals are united to the middle and whose areolation is different. The size is the same.

Hab.—Florissant. Princeton Museum, No. 650.

Porana tenuis, sp. nov.

Calyx large, thin; sepals distinct to the base, oblong, obtuse; veins distinct, distantly obliquely branched.

Resembles *P. macrantha*, Ludw., "Palæontogr." viii. p. 116, pl. xli. fig. 18, but its sepals are still longer—more than 1½ centimeters long, and narrower, half a centimeter. The ramifications of the veins are much more distinct.

Hab.—Florissant. Lacoe's Cabinet, Nos. 65 and 71.

MYRSINEÆ.

MYRSINE, Linn.

Myrsine latifolia, sp nov.

Plate XXXVIII, Fig. 16.

Leaf subcoriaceous, broadly oval or nearly round, truncate at base, very entire; nervation camptodrome.

The leaf, 2 centimeters long and as broad, is broken at the base and the top, and therefore the mode of attachment to the petiole is not seen. The nervation is, however, so much like that of species of this genus that its reference to it seems legitimate. The open, opposite, slightly curving, secondary nerves fork two or three times, and are divided toward the borders, where they abruptly curve and follow close to the margins in short anastomosing bows. The areas between the secondary nerves are obliquely crossed by branching nervilles constituting a loose polygonal areolation.

The affinity of this leaf as to its form and size is with *M. antiqua*, Ung., "Syllog.," p. 20, pl. vii, figs. 7, 7*b*. The European leaf is a little larger and the secondary nerves also a little more curved: the areolation is of the same type. The leaf appears to be unequilateral, and in this and size it is comparable to *M. Chamædrys*, Ung., "Fl. v. Sotzka," p. 42, pl. xxii. figs. 4, 5. The type of nervation of the American species is that of *M. bifaria*, Wall., of India.

The leaf described here is the only one seen as yet of this genus in the North American geological formations; thirty-four species have been described from the European Tertiary. The leaves are generally very small and have probably been unobserved until now.

Hab.—Florissant. Princeton Museum, No. 874.

SAPOTACEÆ.

BUMELIA, Swartz.

The plants of this genus have the leaves alternate, petiolate, coriaceous, and very entire. They inhabit at the present epoch tropical and boreal America. Ten fossil species are described from the European Continent.

Bumelia Florissanti, sp. nov.

Plate XXXIV, Figs. 4, 5.

Leaves thick, obovate, obtuse; lateral nerves thin, at an open angle of divergence, parallel, camptodrome.

The leaves, nearly 5 centimeters long and 3 broad in the upper part, are rounded at the apex, either slightly emarginate or apiculate, gradually narrowed to a very short petiole. Of the nervation nothing is distinct except the thin secondary nerves diverging at base at an angle of 60° to 70°, much curved in passing toward the borders, crossed at right angles by close nervilles, camptodrome. In size and shape these leaves are comparable to *Bumelia subspathulata*, Sap., "Ét.," iii. 3, p. 62, pl. 10, figs. 18–22, and in their different characters to the living *B. retusa* of Jamaica.

Hab.—Florissant; not rare. U. S. Geol. Expl. Dr. *F. V. Hayden*.

DIOSPYROS, Linn.

"U. S. Geol. Rep.," vii, p. 230.

Diospyros brachysepala, Al. Br.

Plate XXXIV, Figs. 1, 2.

Ibid., p. 232, pl. xl. figs. 7–10; lxiii, fig. 6.

The two leaves figured in this volume are more positively identified with the European species than the fragments of "Rep.," vii, pl. xl, whose affinity is still somewhat doubtful on account of the thickness of the secondary nerves.

Hab.—Florissant; not rare. Princeton Museum, Nos. 631, 657, &c.

Diospyros Copeana, Lesqx.

Plate XXXIV, Fig. 3.

"U. S. Geol. Rep.," vii, p. 232, pl. xl, fig. 11.

Though this leaf is shorter and its nervation more distinct, it has evidently the same characters as that described from Elko Station in vol. vii.

Hab.—Florissant. U. S. Geol. Expl. Dr. *F. V. Hayden.*

MACREIGHTIA, A. D. C.

The fossil remains referable to this genus are represented by calyxes. These are merely tripartite; those of *Diospyros* are generally 4 to 6-lobed.

Macreightia crassa, sp. nov.

Plate XXXIV. Figs. 16, 17.

Calyx thick and coriaceous, trilobate; lobes cut to the middle, triangular.

Hab.—Florissant; not rare. Seen in all the collections.

ERICACEÆ.

ANDROMEDA, Linn.

"U. S. Geol. Rep.," vii, p. 234.

Andromeda delicatula, sp. nov.

Plate XXXIV, Figs. 10, 11.

Leaves submembranaceous, not thick, very entire, equally narrowed from the middle upward to a short blunt acumen, downward to a long slender petiole; nervation camptodrome.

These fine leaves average 5 centimeters long and 2 broad in the middle where they are widest. The lateral nerves at an angle of divergence of 40° curve in passing to the borders and follow them in anastomosing bows. They are parallel, unequal in distance; the basilar ones follow close to the borders at a more acute angle of divergence. This and the smaller size of the leaves, more enlarged in the middle, separate this species from *A. protogæa*, Ung., in Heer, "Fl. Tert. Helv.," p. 8, pl. ci, fig. 26.

There is in Lacoe's Cabinet a number of oblong or linear-lanceolate leaves narrowed to a long petiole, exactly similar to those of *A. protogæa* as figured by Heer, *loc. cit.*, but without trace of nervation. They seem indeed referable to the European species.

Hab.—Randolph Co., Wyoming. U. S. Geol. Expl. Dr. *F. V. Hayden.*

Andromeda rhomboidalis, sp. nov.

Leaves rhomboidal in outline, enlarged in the middle, narrowed downward to a long slender petiole and equally so upward to an obtuse apex: nervation obsolete.

The leaves without the petiole are 3 centimeters long, 18 millimeters broad in the middle; the very slender flexuous petiole is broken 1½ centimeters from the base of the leaf.

Species comparable to *A. tremula*, Heer, "Fl. Tert. Helv.," p. 9, pl. ci, fig. 25. The leaves are, however, more enlarged in the middle.

Hab.—Florissant. Lacoe's Cabinet, No. 70.

VACCINIUM, Linn.

Vaccinium reticulatum?, Al. Br.

"U. S. Geol. Rep.," vii, p. 235, pl. lix, fig. 6.

ARALIACEÆ.

ARALIA, Tourn.

"U. S. Geol. Rep.," vii, p. 235.

Aralia dissecta, sp. nov.

Plate XXXV.

Leaves palmately seven-lobed; primary segments cut to three-fourths of the lamina, oblong-lanceolate, deeply lobate, dentate above; secondary divisions lanceolate, obtusely dentate-lobed; sinuses obtuse; secondary nerves subopposite, thick, pinnately branching; nervation craspedodrome.

Of the seven lobes of this fine leaf three are preserved nearly entire and sufficiently represent its character. The leaf, nearly round or fan-shaped in outline, 19 centimeters long from the top of a very thick petiole to the apex of the medial lobe, is cut into seven primary divisions, all pinnately or bipinnately lobate-dentate: the lobes and teeth oblique, slightly turned up, each entered by one of the secondary or of the tertiary nerves, all the nerves therefore corresponding to one division of the leaves and united by nervilles at right angles. There are no intermediate veins passing up to the base of the lobes as in the large fragments which I have referred to *Myrica* as *M. insignis* and *M. Lessigii* of vol. vii, which have apparently a kind of primary division like this leaf.

This fine species is closely related to *Aralia multifida*, Sap., "Ét.," i, 1, p. 115, pl. xii, fig. 1, from which it differs merely by the primary divisions being regularly pinnately lobed, the lobes also pinnately lobed or deeply dentate, the teeth shorter and more obtuse. Saporta compares his species to *Aralia elegans* of New Grenada, a plant cultivated in gardens, which from the figure given by the author seems like a counterpart of the fossil leaf.

Hab.—Florissant. This splendid specimen is in the Princeton Museum, No. 659.

HEDERA, Linn.

Hedera marginata, sp. nov.

Plate XL, Fig. 8.

Leaf small, coriaceous, nearly round in outline, truncate at base, deeply sharply lobate all around; nervation five-palmate from the base, the nerves directed toward the points of the lobes, united by nervilles at right angles.

I know nothing to which this leaf may be related. In shape and nervation it seems a species of *Hedera* comparable by these characters to *H. prisca*, Sap., "Séz. Fl.," p. 380, pl. x, fig. 1, which, however, is a large leaf with short obtuse teeth.

Hab.—Florissant. U. S. Geol. Expl. Dr. *F. V. Hayden.*

AMPELIDEÆ.

CISSUS, Linn.

Cissus parrotiæfolia, Lesqx.

"U. S. Geol. Rep.," vii, p. 239, pl. xl, figs. 15–17.

AMPELOPSIS, Mich.

Ibid., p. 242.

Ampelopsis tertiaria, Lesqx.

Ibid., p. 242, pl. xliii, fig. 1.

SAXIFRAGEÆ.

WEINMANNIA, Linn.

Leaves simple, ternate, quinate or odd-pinnate; petiole articulate; rachis often alate, rarely entire; secondary nerves thin, camptodrome or craspedodrome.

The leaves which I refer to this genus have been referred by authors either to *Zanthoxylum* or *Celastrus*, or especially to *Rhus*, as I have done in vol. vii. Fine figures of species of *Weinmannia* from specimens obtained by Rev. *Probst* from the Tertiary of Biberack, and communicated to me by Heer, show such a close relation to the leaves described from Florissant that their reference to the same genus cannot be doubted.

CF 12

Weinmannia Haydenii, Lesqx.

Plate XLII, Figs. 1-7.

Rhus Haydenii, Lesqx., "U. S. Geol. Rep.," vii, p. 291, pl. lviii, fig. 12.

Leaves imparipinnate; rachis winged; leaflets opposite or alternate, sessile, membranaceous, narrowly lanceolate, obtusely serrate; nervation pinnate, craspedodrome; nervilles at right angles to the secondary veins, anastomosing in the middle of the areas and forming a small polygonal areolation.

The rachis is winged and nerved; the leaflets are joined to the midrib by their primary nerves, and their borders are continued at base by a narrow margin along the rachis.

Hab.—Florissant. Very abundant; seen in all the collections. The figures are from specimens obtained by the U. S. Geol. Expl. Dr. *F. V. Hayden.*

Weinmannia integrifolia, sp. nov.

Plate XLII, Figs. 8-13.

Leaves narrower than in the preceding species; leaflets narrow, entire, oblong or sublinear, blunt at the apex, more distinctly turned upward; nervation camptodrome.

Except that the leaflets are narrower and entire and the nervation consequently camptodrome, the characters are the same and this form may represent only a distinct variety. The leaves of these two species are polyphyllous, the number of their leaflets being much greater than in any other species living at this epoch. This difference and the nearly linear wing of the petiole relate them to *Rhus.*

Hab.—With the preceding and quite as common.

Weinmannia obtusifolia, sp. nov.

Plate XLI, Figs. 4-10.

Leaflets close, the upper pairs decurrent and connate at base, the lower more distant, bordering the rachis by their decurrent base; wing obtusely dentate or convex in the middle; leaflets oblong-obtuse or subspatulate, very entire, more rigid than in the two preceding species, membranaceous; nervation camptodrome.

As in the other species, the leaflets are alternate or opposite, narrowed toward the base or larger toward the obtuse or rounded apex; the leaves are generally smaller, shorter, with fewer leaflets.

Hab.—Florissant; not as frequent as the two preceding ones.

MALVACEÆ.

STERCULIA, Linn.

Schimper remarks, on the present distribution of this genus, that it has made its appearance in Europe at the first stage of the Tertiary, as it is already reported in the "Flora of Sézanne;" that it has had its largest representation in the Miocene, and has since totally disappeared from the continent. The numerous forms of leaves of this genus described in this volume from the Dakota Group prove that the origin of these plants should be removed to the Cretaceous for the American continent at least. The genus is thence found in the divers stages of the Tertiary, but far less frequently here than in Europe.

Sterculia rigida, sp. nov.

Plate XXXIV, Fig. 12.

Leaf subcoriaceous, rigid, cuneate at base, tripalmately lobed; lobes cut to near the base, narrowly lanceolate, sharply acuminate, very entire, the lateral shorter and narrower; nervation obsolete.

I have seen another leaf of the same character since the first was figured, but it does not show anything more except the base, which is cuneate, or like a continuation of fig. 12, to the top of the petiole. The leaves are small, 5½ centimeters between the points of the lateral lobes, 7 centimeters long from the base to the apex of the medial lobe which is 6 centimeters long, the lateral only four. The only species related to this is *S. Labrusca*, common in the Miocene, but the relation is distant.

Hab.—Florissant; very rare. Princeton Museum, No. 667. Lacoe's Collection. No. 44.

TILIACEÆ.

TILIA, Linn.

Tilia populifolia, sp. nov.

Plate XXXIV, Figs. 8, 9.

Leaves large, round or subcordate at base, deltoid-acuminate to the apex, deeply regularly serrate, palmately five-nerved; upper lateral nerves somewhat thicker and more distant, the secondary parallel, slightly curving, branching near the borders. Leaves large, variable in size.

At first the leaf, fig. 8, seems to represent a *Populus* on account of the lateral primary nerves being much stronger than the secondary; but all the nerves and their divisions are craspedodrome; the nervation is positively that of a *Tilia.* In fig. 9 the primary nerves, though more distant, are not stronger, and the teeth of the borders are triangular, somewhat unequal, not turned up as in fig. 8, except toward the base, where they have evidently the same character in both leaves. The teeth are very variable on the borders of the leaves of *Tilia*, even on those of the same tree, and the habitat being the same I refer these to the same species.

Hab.—Florissant. Princeton Museum. Nos. 886 and 887.

ACERACEÆ.

ACER, Linn.

"U. S. Geol. Rep.," vii, p. 260.

Acer æquidentatum, Lesqx.

Ibid., p. 262, pl. xlviii, figs. 1-3.

Acer indivisum, sp. nov.

Plate XXXVI. Figs. 6, 9.

Leaves small, of thin texture, round-truncate in outline, five-nerved and five-lobed; lobes entire, sharply acuminate; sinuses broad, entire or dentate in the middle; petiole comparatively long, inflated under the point of attachment.

The leaves are 5½ centimeters broad between the points of the upper lobes and only 4 centimeters long from the top of the petiole, which is 5½ centimeters long. They are truncate at base, the lower lobes shorter, turned outside at right angles to the medial nerve; the upper lateral ones a little longer, also turned outside. The primary nerves are thin; no trace of secondary nervation is seen.

This species is comparable to *Acer Sibiricum,* Heer, "Fl. Foss. Arct.," v, p. 46, pl. x, figs. 4*b*, 5*a*, 5*b;* xi, fig. 2, differing by the base of the leaves being truncate and entire, not dentate, the sharply acuminate longer lobes, the terminal also entire, the medial nerve being simple like the lateral ones, without branches going to the borders. The affinity of this leaf is more evidently marked with *Acer rubrum*, to which the fruit, fig. 9, is still more intimately related.

Hab.—Randolph Co., Wyoming. U. S. Geol. Expl. Dr. *F. V. Hayden.*

Acer, species.

Plate XXXVI, Figs. 7, 8.

Leaves rounded to the petiole, palmately three-nerved and three-lobate; borders dentate.

The leaves are too much broken for determination and definitive description; they appear related to some of the varieties of *Acer trilobatum*, Al. Br.

Hab.—Florissant.　U. S. Geol. Expl. Dr. *F. V. Hayden.*

SAPINDACEÆ.

SAPINDUS, Linn.

"U. S. Geol. Rep.," vii, p. 263.

Sapindus stellariæfolius, Lesqx.

Ibid., p. 264, pl. xlix, fig. 1.

Sapindus angustifolius, Lesqx.

Plate XXXVII, Figs. 1–8; XXXIX, Fig. 12.

Ibid., p. 265, pl. xlix, figs. 2–7.

The numerous forms figured of this species, common at Florissant, shows the great variety of its leaflets. Though comparatively large, the leaves of pl. xxxix, fig. 12, appear referable to it. The specimens, however, may represent two specific forms, which can be separated only when the nervation is known.

Sapindus coriaceus, Lesqx.

"U. S. Geol. Rep.," vii, p. 265, pl. xlix, figs. 12–14.

Sapindus Dentoni, Lesqx.

Ibid., p. 265, pl. lxiv, figs. 2–4.

Sapindus obtusifolius, Lesqx.

Ibid., p. 266, pl. xlix, figs. 8–11.

There is a fine specimen of this species from Florissant in M. Lacoe's cabinet, No. 48.　The leaflets are disposed as in fig. 8, *l. c.*, but they are still smaller, the lower 1½ centimeters, the upper 1 centimeter, all more distinctly obtuse.

Sapindus inflexus, sp. nov.

Plate XXXII, Fig. 2.

Leaves subcoriaceous, unequilateral at the narrowed base, lanceolate-acuminate; lateral nerves much curved and following the borders in anastomosing with the upper ones.

The form of the leaflet and its nervation indicate its reference to this genus. It is distantly related to *S. undulatus*, Heer, "Fl. Tert. Helv.," iii, p. 62, pl. cxxi, figs. 3–7.

Hab.—Florissant. Princeton Museum, No. 763.

Sapindus lancifolius, sp. nov.

Plate XXXII, Figs. 3–6; XXXVII, Fig. 9.

Leaves subcoriaceous or membranaceous, petioled and more or less unequilateral at the rounded base, lanceolate, long-acuminate, very entire; secondary nerves close, parallel, nearly at right angles to the narrow midrib, straight or slightly curved in traversing the lamina, abruptly curving near the borders and anastomosing in simple bows.

These leaflets, 6½ to 7 centimeters long and more or less than 2 centimeters broad, have the lateral veins close, parallel, united by oblique simple nervilles and nearly without branches. They are distinctly related to *S. Græcus*, Ung., "Fl. v. Kumi," p. 49, pl. xii, figs. 1–23. In this species the veins are equally close and numerous at right angles to the midrib and the leaves have the same form; they are, however, generally smaller. As in those of Florissant, the petiole is 1 centimeter long. In fig. 9 of pl. xxxvii the leaf is narrowed to the petiole, which appears longer; the veins are not as open nor as numerous; its reference to this species is not certain.

Hab.—Florissant. Princeton Museum, Nos. 644 and 645.

DODONÆA, Linn.

I have referred to this genus the seed, pl. xxxvi, fig. 5, on account of its great likeness to that of *D. canescens*, D. C., figured by Ettinghausen in "Fl. v. Här.," pl. xxiii, o. The nucleus is, however, harder, more distinct, and the wings also more distantly veined. It is, perhaps, a seed of *Ulmus*, like those figured, pl. xxvii, fig. 8, from which it differs merely by its slender pedicel. No leaves of *Dodonæa* have been observed in the Green

River Group. The leaves of *Ulmus* are on the contrary very abundant at Florissant and other localities of the North American Tertiary where fossil plants have been obtained.

STAPHYLEACEÆ.

STAPHYLEA. Linn.

"U. S. Geol. Rep.," vii, p. 267.

Staphylea acuminata, Lesqx.

Plate XXXVI, Figs. 1–4.

Ibid., p. 267, pl. xlviii, figs. 4, 5.

The species is not rare at Florissant, but generally the leaves are defaced by maceration and their characters obscurely defined.

FRANGULACEÆ.

EVONYMUS, Tourn.

Leaves opposite, petiolate, ovate, serrate or dentate, pinnately nerved; secondary nerves camptodrome or effaced in the reticulation toward the borders.

Ten fossil species of this genus are described from the European Tertiary, mostly from the Miocene.

Evonymus flexifolius, sp. nov.

Plate XXXVIII, Fig. 13.

Leaves large, ovate-acuminate from an oval base, flexuous at the apex, narrowed from the middle to the petiole, sharply deeply serrate; secondary nerves alternate, equidistant and parallel, camptodrome.

The leaf without the petiole is 16½ centimeters long, 5 centimeters broad in the middle, where it is oval-oblong, narrowed upward to a long flexuous acumen and more rapidly to the petiole, which is 3 centimeters long. The teeth of the borders are turned upward, equal, becoming short toward the acumen, deeply cut; the nervation is truly camptodrome, the veins being effaced near the borders and not entering the teeth directly as it is incorrectly figured.

This leaf has the characters of *Evonymus Proserpinæ*, Ett., "Bil. Fl.," iii, p. 30, pl. xlviii, figs. 6, 7. It is of the same size and shape, more grad-

nally and longer acuminate; the border teeth are larger and more acute.
The details of nervation are obsolete.

Hab.—Randolph Co., Wyoming. U. S. Geol. Expl. Dr. *F. V. Hayden.*

CELASTRUS, Linn.

"U. S. Geol. Rep.," vii, p. 268.

Celastrus Lacoei, sp. nov.

Leaves subcoriaceous, obovate or spatulate, rounded and dentate at the apex.

The leaf is remarkably similar in character to those described by Heer
as *C. cassinefolius*, Ung., in "Fl. Tert. Helv.," iii, p. 67, pl. cxxi, figs. 24–26,
whose leaves are longer and narrower, obtusely dentate or rather crenulate from the middle upward.

Hab.—Florissant. Lacoe Collection, No. 49.

Calastrus Greithianus, Heer.

"Fl. Tert. Helv.," iii, p. 70, pl. cxxi, fig. 63.

Leaves small, broadly oval, obtuse, very entire, abruptly narrowed to the petiole;
lateral nerves nearly at right angles to the midrib, camptodrome.

Two leaves from Florissant are referred to this species. One is of
the same size, form, and nervation as that figured by Heer, the other is
more gradually narrowed to the base, lacerated at the rounded apex. This
last leaf is more like *C. Bruckmanni.* Heer, *l. c.*, fig. 32.

Hab.—Florissant. Lacoe Collection, No. 74.

Celastrus fraxinifolius, sp. nov.

Plate XXXIII, Figs. 2–4; Plate XL, Fig. 10.

Leaves membranaceous, narrowly elliptical in the middle, lanceolate, acuminate,
blunt at the apex, narrowed and decurrent to the petiole, crenulate-dentate; secondary
nerves at an acute angle of divergence, curving to the borders and reticulate along
them.

The leaves, 5 to 7 centimeters long, averaging 2 centimeters in width
in the middle, are mostly equilateral at the narrowly cuneate base, short-
petioled, the petiole ½ centimeter long, being bordered by the decurrent
base of the leaves; the lateral nerves unequally distant, much and
unequally curved in traversing the lamina, follow the borders in multiple
reticulations without entering the teeth, which are distant, obtuse, sometimes obsolete.

The leaves have a great affinity in their characters to those of species of *Fraxinus*. They are, however, equilateral on the borders and the nervation is different. Figure 3 of pl. xl may represent another species; the leaf is broader and slightly unequilateral. The decurrent base of the leaf and the type of nervation are the same.

Hab.—Florissant; not rare. U. S. Geol. Expl. Dr. *F. V. Hayden*. Fig. 10 represents two leaves, Nos. 648 and 870 of the Princeton Museum.

Celastrinites elegans, sp. nov.

Plate XXXI, Figs. 9, 10.

Leaves nearly round, membranaceous, somewhat long-petioled, crenate on the borders; nervation pinnate; secondary veins oblique, parallel, reticulate and effaced along the borders.

The leaves are very small, 1½ to 2½ centimeters long and about the same width, rounded or broadly cuneate to the petiole.

Figure 10 is truncate at base and its nervation appears triple-nerved, as in *Populus;* but the surface is somewhat erased and the upper secondary nerve obsolete, and as all the other characters are alike the difference is not considered.

Hab.—Florissant. Princeton Museum, Nos. 799 and 868.

ILICEÆ.

ILEX, Linn.

"U. S. Geol. Rep.," vii, p. 269.

Ilex pseudo-stenophylla, sp. nov.

I. stenophylla, Lesqx.; Hayden's "Ann. Rep.," 1871, Supp't, p. 8.

Leaves small, coriaceous, very entire, obovate or oblanceolate, obtuse, short-pedicellate; medial nerve thin; lateral nerves very oblique, much curved near the borders, anastomosing.

The leaf is much like those of *I. stenophylla*, Ung., "Syllog.," ii, p. 14, pl. iii, figs. 15, 27, being, however, smaller with a shorter broad pedicel. The nervation is like that of figs. 24 and 25 of Unger. The leaves described in Hayden's "Ann. Rep.," *loc. cit.*, have the same degree of affinity to Unger's species and are all larger. They apparently represent an American variety of the species.

Hab.—Florissant. No. 59 of Lacoe's Collection.

Ilex microphylla, sp. nov.

Leaves small, coriaceous, obovate or spatulate, rounded and denticulate at the apex, narrowed to a short broad petiole; secondary nervation obsolete.

The leaf, 2½ centimeters long, 7 millimeters broad in the upper part, is gradually narrowed to a petiole 7 millimeters long. Its affinity, which is close indeed, is with *Ilex ambigua*, Ung., "Syllog.," ii. p. 14, pl. iii, fig. 29, from which it differs merely by the gradually narrowed base of the leaf and the longer petiole.

Hab.—Florissant. No. 60 of Lacoe's Collection.

Ilex maculata, sp. nov.

Plate XLIV, Fig. 5.

Leaves coriaceous, obovate, obscurely and irregularly crenulate, narrowed to the petiole; medial nerve narrow, the lateral at a broad angle of divergence, a little curved in traversing the blade, effaced toward the borders.

The leaf is badly preserved; its surface is maculated or gnawed by parasite hypophylles or insects. Its shape and thick consistence appear to refer it to this genus.

Hab.—Alkali Station. Professor *Scudder*.

Ilex Wyomingiana, Lesqx.

"U. S. Geol. Rep.," vii, p. 270, pl. 1, fig. 1.

Ilex? affinis, Lesqx.

Ibid., p. 270, pl. 1, figs. 2, 3.

Ilex subdenticulata, Lesqx.

Ibid., p. 271, pl. 1, figs. 5, 6–6*b*.

Ilex dissimilis, Lesqx.

Ibid., p. 271, pl. 1, figs. 7–9.

Ilex quercifolia, sp. nov.

Plate XXXVIII, Figs. 2–5.

Leaves coriaceous, short-petioled, obovate, abruptly acuminate, irregularly acutely dentate from near the base; secondary nerves at a broad angle of divergence, slightly curved in passing to the borders, entering the teeth directly or by branchlets; intermediate tertiary veins short, anastomosing with nervilles in the middle of the areas.

The leaves are very variable in size (from 12 millimeters long to nearly 6 centimeters, and 5 millimeters to 2 centimeters broad); the petiole is thick and short (6 millimeters long); the teeth turned outside, sharply pointed, are distant and variable in length, separated by obtuse sinuses; the acumen is sharply pointed.

The relation of this species is distinctly indicated to *Ilex dryandræfolia*, Sap., "Ét.," i, 2, p. 89, pl. x, fig. 8, a leaf which is much like fig. 2 of our plate, and which merely differs by the secondary nerves being at right angles to the midrib, rather curved backward than upward, a difference scarcely noticeable enough to authorize specific distinction. The *Ilex odora*, Sieb. and Zucch., of Japan, has the leaves remarkably similar to both these fossil species.

Hab.—Florissant. U. S. Geol. Expl. Dr. *F. V. Hayden.*

Ilex grandifolia, sp. nov.

Plate XXXVIII, Fig. 1.

Leaves large, membranaceous, oblanceolate or obovate, irregularly dentate; lateral nerves very oblique, more or less curved in traversing the blade, camptodrome, joined to the borders and the teeth by anastomosing nervilles.

The leaf seems to have been very large, the fragment preserved (the upper half) being 8 centimeters long and 5 centimeters broad. It appears to have been rounded at the apex and gradually narrowed to the base, the lower lateral nerves being very oblique and following the borders in curves. The nervation is irregular. The lateral nerves, diverging about 30°, are distant, parallel, with few intermediate tertiary shorter thin veins, and in their curves they generally ascend to near the borders, but also sometimes curve in the middle of the areas, anastomosing with the divisions of the first nerves above and sending strong outside branches toward the borders. The teeth are somewhat unequal but not as large as in the preceding species, more or less inclined upward, acute. The subdivision of the primary areas is by nervilles at right angles to the nerves, anastomosing generally at right angles with the thinner tertiary veins, producing a large irregularly quadrate areolation.

Hab.—Florissant. U. S. Geol. Expl. Dr. *F. V. Hayden.*

Ilex knightiæfolia, sp. nov.

Plate XL., Figs. 4, 5.

Leaves membranaceous, linear in outline, decurrent to the petiole, rounded and acuminate at the apex, deeply dentate; secondary nerves at right angles, curving abruptly and anastomosing at right angles at a distance from the borders, joined to the teeth by nervilles; teeth large, irregular in distance, turned outside and sharply pointed.

These leaves have peculiar characters which seem to refer them to some types of the *Proteaceæ* of New Holland, *Banksia Hugelii*, R. Br., and species of *Knightia*. The small leaf, fig. 5, is better preserved but not sufficiently so to show the base of the leaf which, being lacerated, appears to follow and border the thick petiole to its point of attachment. The teeth, like the secondary nerves, are at right angles to the midrib except near the apex, which is formed of a sharply angular point; the secondary nerves are separated by slightly thinner and shorter tertiary ones, anastomosing with nervilles at right angles in traversing the areas and united to the upper part by curves or strong nervilles also at right angles.

Hab.—Florissant.   U. S. Geol. Expl. Dr. *F. V. Hayden.*

RHAMNEÆ.

"U. S. Geol. Rep.," vii, p. 272.

PALIURUS, Tourn.

Paliurus Florissanti, Lesqx.

Ibid., p. 274, pl. 1, fig. 18.

Paliurus orbiculatus, Sap.

Plate XXXVIII, Fig. 12.

Saporta, "Ét.," iii, 2, p. 182, pl. vii, fig. 6.

Leaves small, membranaceous, orbicular, very entire, triple-nerved from the base; lateral nerves curved upward in ascending to near the apex, where they unite to the secondary nerves which are distant and few.

Though the nervation is not as distinct as in the leaf published by Saporta, the affinity is so clear that it is not possible to doubt specific identity; the basilar nerves, equally branching, ascend high, joining the few secondary nerves, one of which only is distinct in the specimen of Florissant and two only on that figured by Saporta, who described the tertiary veinlets as flexuous and reticulate. The leaf is nearly of the same size, 2 centimeters in diameter both ways.

Hab.—Florissant. U. S. Geol. Expl. Dr. *F. V. Hayden.*

ZIZYPHUS, Mill.

"U. S. Geol. Rep.," vii, p. 275.

Zizyphus cinnamomoides, Lesqx.

Ibid., p. 277, pl. lii, figs. 7, 8.

RHAMNUS, Linn.

Ibid., p. 278.

Rhamnus oleæfolius, sp. nov.

Plate XXXVIII, Fig. 14.

Leaves thick, oblong-lanceolate, narrowed at base, blunt at the apex; secondary veins thick, at an acute angle of divergence, curving close to the borders.

The leaf, 6½ centimeters long, 18 millimeters broad, has the primary and secondary nerves thick, but no trace of nervilles; the lateral veins are nearly straight to near the borders and abruptly curve in reaching them, appearing to join the margin by their ends. The same character of nervation is remarked in *R. marginatus*, Lesqx., "Trans. Phil. Soc.," vol. xiii, p. 420, pl. xxii, figs. 3–5, which, however, differs much in the form and size of the leaves.

Hab.—Florissant. Princeton Museum, No. 687.

Rhamnus notatus?, Sap.

Plate XXXVIII, Fig. 15.

Sap., "Ét.," iii, 1, p. 108, pl. xi, fig. 5.

Leaves subcoriaceous, very short-petioled, entire or slightly undulate in the upper part, round ovate, obtusely pointed; lateral nerves 6 to 7 pairs, parallel, curved; nervilles oblique, transversely reticulate.

This leaf is, in its form and size, like a counterpart of that of Saporta, *l. c.* It is also rounded at base to a very short petiole, curved toward the apex and there obscurely undulate or crenulate. The lower secondary veins are opposite, three pairs. In the figure of the French author all the veins are alternate except the basilar ones; but there is also no trace of nervilles visible as upon the specimens of Florissant.

Hab.—Florissant. Princeton Museum, No. 643.

TEREBINTHINEÆ.

JUGLANDEÆ.

"U. S. Geol. Rep.," vii, p. 283.

JUGLANS, Linn.

Juglans Schimperi, Lesqx.

Ibid., p. 287, pl. lvi, figs. 5–10.

Juglans denticulata, Heer.

Ibid., p. 289. pl. lviii, fig. 1.

Juglans Florissanti, sp. nov.

Leaf large, lanceolate-acuminate from a rounded unequilateral base; lateral veins thick, much curved in traversing the blade, camptodrome; borders dentate.

The leaf is 11 centimeters long, 4½ centimeters broad in the middle; its surface is rough and altogether of coarse aspect—the primary and secondary nerves being thick. The details of areolation and subdivisions of the nerves are obsolete. It is comparable to a leaf of *J. bilinica*, figured in Heer, "Fl. Tert. Helv.," p. 90, pl. cxxx, fig. 7, but it is thicker, coarser, with more prominent nerves.

Hab.—Florissant. Lacoe's Collection, No. 80.

Juglans alkalina, Lesqx.

"U. S. Geol. Rep.," vii, p. 288, pl. lxii. figs. 6–9.

Juglans costata, Ung.

Plate XXXIX, Fig. 5.

Carya costata, Ung., "Syllog.," p. 41, pl. xxxix, fig. 16.
Juglans costata, Ludw., "Palæontogr.," viii, p. 135, pl. lvii, fig. 7 (leaf); liv, fig. 15 (nut).
Juglans acuminata?, Heer, Lesqx., Suppl. to Hayden's "Ann. Rep.," 1871, p. 8.

Leaflets broadly oval, obtuse, slightly mucronate, somewhat unequilateral or turned to one side, rounded at base to a short petiole; nervation camptodrome. Nut round-ovate, short-pointed; lobes of the seed simple, oblong.

In the short description of the leaflet as *J. acuminata?*, *loc. cit.*, I remarked that it has exactly the same characters as the one figured by Heer, "Fl. Tert. Helv.," pl. cxxix, fig. 6, which appears far different from any other forms of this species, and that it is comparable to *J. costata*, Ung., as figured by Ludwig, *l. c.* As one of the specimens of Florissant has a nut very much like that published by the same author, *l. c.*, the

identification of the American specimens with Ludwig's species is legitimate.

Hab.—Florissant. Princeton Museum. No. 712 (nut).

CARYA. Nutt.

Carya bilinica, Ung.

Plate XXXIX, Figs. 1, 2, 13.

Ung., "Syllog.." p. 39, pl. xvii, figs. 1–10; "Fl. v. Kumi.." p. 54, pl. xiv, fig. 13; Ett., "Bil. Fl.," iii, p. 46, pl. li, figs. 4–6, 13, 15; lii, figs. 3, 4, 7–11.

Leaves odd-pinnate; leaflets short-petioled, oblong or narrowly ovate, lanceolate, acuminate, serrate; lateral nerves camptodrome, parallel.

These fine leaves correspond to the description and figures given of the species by European authors; the borders of the leaves are more or less distinctly serrulate, as shown in fig. 2; fig. 13 shows a variety represented also by the specimens of Mr. Lacoe, which might, perhaps, be separated into a different species, but except the smaller size of the long-acuminate leaflets, the characters are the same.

Hab.—Florissant; not rare. U. S. Geol. Expl. Dr. *F. V. Hayden.* Lacoe's Collection, No. 40, in leaves still smaller than fig. 1.

Carya rostrata, (Goepp.), Schp.

Plate XXXIX, Fig. 4.

Ludw., "Palæontogr.," viii, p. 136, pl. lv, figs. 5–7.

I refer this nut to the species of Ludwig described as quoted above. As we have only on the Florissant shale the representative of a drupe or of the husk, its reference to the European species known by fruits and leaves is not more ascertainable than that of the preceding.

Hab.—Florissant. Princeton Museum, No. 711.

Carya Bruckmanni?, Heer.

Plate XXXIX, Fig. 6.

Heer, "Fl. Ter. Helv.," iii, p. 93, pl. cxxvii, fig. 52.

Fruits small, oval, constricted into an obtuse apex, costate.

The fruit is still smaller than that in Heer, *loc. cit.*, and as the inside of the nut only is shown upon the face of the specimen it is not possible to see whether this small nut is costate. Therefore, as in the two preceding species, the reference is uncertain.

Hab.—Florissant. Princeton Museum. No. 709.

PTEROCARYA, Kunth.

Pterocarya Americana, Lesqx

"U. S. Geol. Rep.," vii, p. 290, pl. lviii, fig. 3.

ENGELHARDTIA, Leschen.

Leaves abruptly pinnate; leaflets unequilateral, generally resinose, punctate on the lower surface; primary nerves strong, secondary thin, camptodrome, anastomosing. Flowers agglomerated in paniculate ears; drupe small, connate at base to a tri-alate involucre; dorsal lobe generally absent (in fossil specimens), epicarp coriaceous, putamen bicostate.

Engelhardtia oxyptera, Sap.

"Ét.," ii, p. 314, pl. xii, fig. 2.

Lobes of the involucre linear-oblong, obtusely pointed, the lateral half as long as the middle; medial nerve distinct to the point, the lateral open-oblique, camptodrome.

The involucre from the base of the nucleus to the top of the medial lobe is 3 centimeters long, a little more than 2 to the top of the lateral ones. The basilar nervation of the middle lobe is in two short basilar parallel nerves and above in curved secondary nerves, as in the lateral lobes; all the nerves are camptodrome and anastomosing. The involucre is only slightly larger than in Saporta's figure; the nervation is the same.

Hab.—Florissant. *Wm. Cleburne.*

ANACARDIACEÆ.

RHUS, Linn.

"U. S. Geol. Rep.," vii, p. 201.

Rhus fraterna, sp. nov.

Plate XLI, Figs. 1, 2.

Leaves simple, submembranaceous, long-petioled, rhomboidal-oval, equally narrowed to the acute apex and to the petiole, very entire; medial nerves narrow, the lateral thin, nearly parallel, oblique, much branching, and obliquely reticulate toward the borders.

The leaves average 4 centimeters long and 2 broad in the middle, the widest part. The nervation is delicate but very distinct; the secondary nerves, at an angle of divergence of about 40°, pass toward the borders, slightly curved and obliquely branching, especially near the borders; the nervilles are mostly at right angles to the midrib. Except that the petiole of the leaves is longer, nearly 2 centimeters, and the leaves slightly more enlarged in the middle, the species is, in all its characters, identical with *Rhus palæocotinus*, Sap., "Ét.," ii, p. 352, pl. xii, fig. 6, closely allied to the well-known *R. Cotinus*, Linn.

Hab.—Florissant. Princeton Museum, Nos. 783 and 875.

Rhus coriarioides, sp. nov.

Plate XLI, Fig. 3.

Leaves odd-pinnate; leaflets narrowly lanceolate, gradually acuminate, narrowed in rounding to the base, sessile; borders distantly serrate; lateral nerves curved, craspedodrome.

The leaflets are opposite, at least in the upper part of the leaves, 6½ centimeters long, 10 to 12 millimeters broad toward the base; the teeth are short, turned upward, gradually smaller toward the apex, where the borders are entire as near the base. The affinity of this species is with *Rhus glabra*, Linn., of the present North American Flora, and especially with the European *R. coriaria*, Linn., which merely differs by the larger teeth of the borders.

Hab.—Florissant. Princeton Museum, No. 858.

Rhus cassioides, sp. nov.

Plate XLI, Fig. 11.

Leaves trifoliate or odd-pinnate; leaflets obovate, the terminal twice as large as the lateral ones, entire; lateral veins close, 8 to 10 pairs, parallel, curved in passing to the borders, craspedodrome.

The specimen does not indicate whether the three leaflets figured pertain to an odd-pinnate leaf or to a trifoliate one, the axis or pedicel being broken under the point of attachment of the leaflets. The terminal one is 2½ centimeters long, 12 millimeters broad above the middle; the lateral 14 to 15 millimeters long and 6 millimeters broad; the lateral veins, quite distinct, follow close to the borders in their curves and are united by close nervilles at right angles, simple or anastomosing in the middle.

c r 13

The nervation is like that of some species of *Cassia*—*C. lignitum, C. ambigua,* Ung., for example.

Hab.—Florissant.   U. S. Geol. Expl. Dr. *F. V. Hayden.*

Rhus Hillae, sp. nov.

Plate XLI, Figs. 12-15.

Leaves irregularly pinnately divided; terminal leaflets large, pyramidal, more or less rapidly narrowed to the base, deeply irregularly dentate; lateral pinnules small, nearly at right angles, ovate, acute, dentate, alternate or opposite, subdecurrent, sessile.

These leaves, which seem to have been compound and odd-pinnate, are represented in the fossil state merely by the terminal pinnules and one or two of the lateral ones attached to one side of their base, figs. 13, 14, or one pair opposite and sessile on the rachis at a distance from the terminal pinnule, fig. 12.   The nervation is distinct.   As seen in fig. 13, the secondary nerves are very oblique, straight, with intermediate shorter tertiary veins and nervilles at right angles.

The species is comparable to *Rhus incisa,* Sap., "Ét.," iii, 1. p. 111. pl. ii, fig. 4, which is made of a single small leaflet similar to fig. 15 of our plate.

Hab.—Florissant.   Fragments and pinnules of this species have been seen in all the collections made by Mrs. *Hill.*

Rhus acuminata, Lesqx.

Plate XLII, Figs. 14-17.

Lesqx., Suppl. to Hayden's "Ann. Rep.," 1874, p. 8.

Leaflets narrowly ovate, lanceolate, acuminate; borders deeply dentate from near the base; lateral nerves open, joining the midrib nearly at right angles, much curved, craspedodrome.

These leaflets have great analogy of character with the terminal leaflets of *Weinmannia* as seen in pl. xlii, fig. 3.   They cannot be referred to this genus, however, as they are contracted at base to a narrow not winged petiole.   Their relationship also, considering them as mere leaflets either terminal or lateral, is with the preceding species, being by their shape, the teeth of the borders and the nervation, intermediate between this and the following species. The secondary veins are close, parallel, with intermediate shorter tertiary veins of the same character as in *R. Hilliæ*.

Hab.—Florissant. U. S. Geol. Expl. Dr. *F. V. Hayden.*

The specimen described in Suppl. to "Annual Report," 1871, is from Green River.

Rhus subrhomboidalis, sp. nov.

Plate XLI, Figs. 16–19.

Leaflets membranaceous, ovate or sub-rhomboidal, rounded to a short petiole, deeply dentate, acuminate; lateral nerves curved, craspedodrome.

Though these three leaflets are so much alike in their forms that it is not possible to refer them to two species, their nervation is very different on account of the position of the large teeth, one or two on each side. In fig. 19 the teeth are in the upper part of the leaflet and the lateral veins curve upward to reach them, and are distant from the upper more open parallel ones; in the other leaflets, figs. 17 and 18, the two pairs of teeth being lower, the lateral nerves are merely curved in their direction toward them and parallel from the base. It is not possible to decide whether these leaflets pertain to pinnate or to trifoliate leaves, like those of the common and so very variable *R. aromatica.* Their relation to those described by Saporta as *R. rhomboidalis,* "Ét.," iii, 111, p. 206, pl. xvi, figs. 2, 3, is remarkably close.

Hab.—Florissant. U. S. Geol. Expl. Dr. *F. V. Hayden,* and also in the Collection of the Princeton Museum, Nos. 751 and 832.

Rhus vexans, sp. nov.

Plate XLI, Fig. 20.

Leaves trifoliate, long-petioled; leaflets cuneiform, enlarged, obtusely dentate or lobate in the upper part and there abruptly narrowed to an obtuse apex; nervation mixed.

This small leaf is so exactly similar to a variety of *R. aromatica* (*R. trilobata,* Nutt.), especially found living in Texas, that it is scarcely possible to find any point of difference. In the living species the terminal upper lobes of the pinnules are more distinctly dentate, but its smaller leaves, of the same size as the one figured, have exactly the same subdivisions. The nervation is also the same, the lower lateral veins being camplodrome, the upper entering the teeth or lobes, all being obliquely short branched. The relationship is also marked with the preceding species, which evidently pertains to that peculiar and variable type of *R. aromatica* which is still universally distributed in innumerable varieties through the North American continent from the 30° to the 43° of latitude.

Hab.—Florissant. Princeton Museum. No. 718.

Rhus trifolioides, sp nov.

Leaves trilobate; leaflets oval; the medial slightly obovate and a little longer, narrowed to a short petiole; the lateral sessile, all apiculate and dentate to the middle.

The medial leaflet is 2½ centimeters long, 12 millimeters broad in the middle, the lateral ones 2 centimeters long and 1 broad, not as distinctly dentate as the middle. The teeth are sharp, turned exactly to the outside. The leaf is comparable to *R. Napaeorum*, Ung., "Syllog.," i, p. 43, pl. xx. fig. 11, differing by the form of the oval sharply dentate leaflets. The pedicel is broken 1 centimeter below the base of the leaflets, the nervation indistinct.

Hab.—Florissant. Lacoe's Collection, No. 58.

Rhus rosæfolia, Lesqx.

"U. S. Geol. Rep.," vii, p. 293, pl. xlii, figs. 7–9.

ZANTHOXYLEÆ.

ZANTHOXYLON, Linn.

Zanthoxylon spiræfolium, sp. nov.

Plate XL. Figs. 1–3.

Leaves odd-pinnate; leaflets ovate, acute, or blunt at the apex, obscurely serrate, short-petioled; secondary nerves at an acute angle of divergence, parallel, simple or forking, camptodrome.

The leaflets vary from 1½ to 2½ centimeters long and from 7 to 14 millimeters broad; the lateral nerves appear craspedodrome in fig. 1. But in figs. 2, 3, where the veins are more distinct, they are seen joined to the teeth by nervilles and camptodrome.

This species is closely allied to *Z. juglandinum* and *Z. serratum*, Heer, represented "Fl. Tert. Helv.," pl. cliv, figs. 36 and 37. Upon the leaf, fig. 2, there is a small fruit of *Sapindus* (enlarged, fig. 2*a*), comparable to that of *S. rubiginosus*, figured in Ung., "Syllog.," i, p. 34, pl. xv, fig. 10.

Hab.—Florissant. U. S. Geol. Expl. Dr. *F. V. Hayden*.

AILANTHUS, Desf.

"U. S. Geol Rep.," vii, p. 294.

Ailanthus longe-petiolata, sp. nov.

Plate XL, Figs. 6, 7.

Leaflets subcoriaceous, narrowly ovate-lanceolate, gradually acuminate, rounded in narrowing to a long petiole, irregularly obtusely dentate; secondary nerves close, open, curving near the borders or entering the teeth; tertiary nerves thinner, nearly as long as the secondary; nervilles at right angles.

The leaflets, 10 centimeters long and 3 broad in the middle, are a little smaller than those of *Ailanthus driandroides*, Heer, "Fl. Tert. Helv.," pl. cxxvii, fig. 32, which has the same form and an analogous nervation. In the American leaf most of the secondary nerves seem to enter the teeth or to run to the borders; but in the upper part of the leaves, where the borders are more distinct, the nerves are evidently camptodrome. It is a mixed nervation, same as seen upon the leaflet of Heer, *l. c.* The leaf however represents a different species, the teeth being obtuse and the petiole very long, too long for a leaflet of *Ailanthus*, except if it should represent a terminal one. The lower or basilar tooth on the leaflet is protruding outside and apparently glandulose, a peculiar character of *A. glandulosa* so generally cultivated now. Fig. 7 may not represent the fruit of the same species, though I have not seen any other leaf from the same locality which could be referred to this genus. The *samara* is equally winged on both sides of the seed, oblong, obtuse at both ends, slightly constricted in the middle. The fruit has a close affinity to that of *Ailanthus recognita*, Sap., "Ét.," i, p. 105, pl. viii, fig. 7.

Hab.—Randolph Co., Wyoming. U. S. Geol. Expl. Dr. *F. V. Hayden*.

MYRTACEÆ.

"U. S. Geol. Rep.," vii, p. 295.

EUCALYPTUS, Heer.

Eucalyptus Americana, Lesqx.

Ibid., p. 296, pl. lix, figs. 11, 12.

ROSIFLOREÆ.

AMELANCHIER, Medic.

Amelanchier typica, sp. nov.

Plate XL, Fig. 11.

Leaves submembranaceous, petioled, ovate, acute, serrate; nervation campto-drome.

This leaf seems to represent the living *A. Canadensis* in its more common or typical form, differing in nothing except the rounded base of the leaf, which is generally slightly cordate in the living species. I say generally, for some of its leaves are also rounded just as in the fossil form. The leaf, 8 centimeters long, 4 centimeters broad in the middle, has a petiole 2 centimeters long. The nervation is similar, the lateral nerves being only a little more distant. The average number of secondary nerves in leaves of *Amelanchier Canadensis* is 8 to 11, while the fossil leaf has only 9. But often large leaves of the living species have no more than 9.

Hab.—Florissant. Princeton Museum, No. 691.

CRATÆGUS, Linn.

Cratægus acerifolia, sp. nov.

Plate XXXVI, Fig. 10.

Leaf petioled, lanceolate in outline, deeply lobate, irregularly dentate; lobes lanceolate, acuminate; nervation craspedodrome.

The substance of the leaf is thickish, but not coriaceous; the leaf is gradually narrowed to the petiole, single-lobed on one side, the lobe being longer, and twice-lobed on the other side, where the lobes are shorter—all irregularly dentate. The secondary nerves are all craspedodrome, entering the lobes and the teeth; but their divisions, at least near the points of the lobes, are camptodrome, the borders being nearly entire.

This leaf has the facies of an *Acer*. I find nothing in the fossil plants described by authors to which it may be compared.

Hab.—Florissant. Princeton Museum, No. 660.

ROSA. Linn.

Rosa Hilliæ, sp. nov.

Plate XL., Figs 16. 17.

Leaves small; leaflets oval, obtuse or short-pointed, serrate; stipules large, lanceolate, acuminate; nervation camptodrome.

These beautiful small leaves represent this genus more distinctly than any of the other fossil leaves which as yet have been referred to it. The leaflets are rather obtuse, the lateral much smaller, 5 to 15 millimeters long, 3 to 7 millimeters broad—all short-pediceled like the terminal ones; the nervation is camptodrome. the figure shows it mostly craspedodrome, a mistake evidently, for as seen on the left side of the largest pinnule, fig. 16, the veins are curved. The nervation near the borders is not quite distinct on the specimens.

Hab.—Florissant. Princeton Museum, No. 768. Also in the collection of the U. S. Geol. Expl. by Dr. *F. V. Hayden.*

AMYGDALUS. Linn.

Amygdalus gracilis, sp. nov.

Plate XL., Figs. 12-15; XLIV, Fig. 6.

Leaves ovate-lanceolate, gradually narrowed to the acuminate point and in the same degree to the petiole, serrulate; lateral nerves at a more or less acute angle of divergence, much curved, camptodrome and reticulate along the borders.

These fine leaves of solid membranaceous tissue average 7 centimeters long and 2 broad, with a slender petiole about 2 centimeters long. They are more or less distinctly minutely serrate; the nerves, open at base and much curved toward the borders, are joined by undulate nervilles nearly at right angles.

Fig. 6 of pl. xliv is a leaf slightly longer acuminate, with obsolete nervilles, but without any important difference from the normal form.

The leaves are related to *A. pereger*, Ung., in Heer, "Fl. Tert. Helv.," iii, p. 95. pl. cxxxii. figs. 8-12. The fruits, figs. 14 and 15. appear to belong to this genus and possibly to this species. The reference is of course hypothetical.

Hab.—Florissant. U. S. Geol. Expl. Dr. *F. V. Hayden.* Fig. 12 is from a specimen, No. 865. of the Princeton Museum. The specimen, fig. 6. is from Randolph County. Wyoming. Prof. *Scudder.*

LEGUMINOSÆ.

CYTISUS, Linn.

Cytisus modestus, sp. nov.

Plate XXXIX, Figs. 9, 10, 11.

Leaves trifoliate; leaflets sessile, ovate-lanceolate, acute, borders entire; secondary nerves camptodrome.

The small leaves, with leaflets 2 to 3 centimeters long, 5 to 8 millimeters broad, have the nervation mostly obsolete. I do not find them related to any fossil species published. Fig. 9 appears to have the borders serrulate, but that is probably caused by maceration and erosion. It has the same characters.

Hab.—Florissant. U. S. Geol. Expl. Dr. *F. V. Hayden.*

Cytisus Florissantianus, sp. nov.

Plate XXXIX, Fig. 14.

Leaf long-petioled; leaflets entire, ovate-lanceolate, the middle short-pedicellate, the lateral sessile, unequilateral at base; nervation camptodrome.

The leaflets appear acuminate, but the point is broken; they are rounded in narrowing to the base, and the borders are entire, only slightly undulate. This species is scarcely different from *C. Freybergensis*, Ung., "Syllog.," ii, p. 19, pl. iv, fig. 2, from which it merely differs by the leaflets being a little longer and narrower. The nervation is of the same type, and if the leaflets of the American leaf are obtuse the species should be considered as identical.

Hab.—Florissant. U. S. Geol. Expl. Dr. *F. V. Hayden.*

DALBERGIA, Linn. fil.

Dalbergia cuneifolia, Heer.

Plate XXXIV, Figs. 6, 7.

Heer, "Fl. Tert. Helv.," iii, p. 104, pl. cxxxiii, fig. 20.

Leaves pinnate; leaflets sessile, membranaceous, cuneate to the base, emarginate at the apex; secondary nerves thin, at an acute angle of divergence.

The leaves are small, averaging 3 centimeters long, 1½ broad near the middle, from which they are gradually narrowed to the somewhat enlarged point of attachment. The lateral nerves are at an acute angle of divergence of 40° on the right side, a little more open on the left, ascending high and reticulate along the borders; the areolation is formed of nervilles at right angles, forking or anastomosing in the middle of the areas, rarely simple.

These leaves only differ from the one described by Heer under this name in their slightly larger size and in the apex being a little more deeply emarginate. The nervation is peculiar and evidently of the same type as in the European leaves, where the lateral nerves are, however, somewhat obsolete. The secondary nerves, four pairs, are distant, alternate, the upper pairs curving inward toward the apex of the midrib.

Hab.—Florissant. Princeton Museum, Nos. 790, 791.

CERCIS, Linn.

Cercis parvifolia, sp. nov.

Plate XXXI, Figs. 5-7.

Leaves small, membranaceous, round or subtruncate at base, broadly cuneate to the slightly-pointed apex, very entire, five-nerved at base; medial nerve slightly stronger, secondary nerves camptodrome.

The three leaves figured and a few others seen in the shale of Florissant are small comparatively to those of this genus described as fossil. They are equilateral, enlarged, and truncate or subcordate at base; the basilar nerves are at right angles; the lateral at an angle of divergence of 30° to 40° are camptodrome like their divisions. The reticulation is obsolete. None of the few fossil species of this genus are comparable to this. The leaves vary from 1½ to 3 centimeters in width, being as long as broad.

Hab.—Florissant. Princeton Museum, Nos. 766, 767. Figs. 5 and 6; the other from the U. S. Geol. Expl. Dr. *F. V. Hayden.*

PODOGONIUM, Heer.

' U. S. Geol. Rep.," vii, p. 293.

Podogonium acuminatum, sp. nov.

Plate XL, Fig. 9.

Leaflets sessile, subcoriaceous, very entire, oblong, obtusely acuminate, narrowed to a short petiole, slightly unequilateral at base; lateral nerves close together, very open or nearly at right angles to the midrib, curved, camptodrome; tertiary nerves parallel, as long as the secondary, thin.

The small leaflet, a little more than 4 centimeters long and 1 broad, has the peculiar nervation of species of this genus, especially like that of *P. latifolium*, Heer, "Fl. Tert. Helv.," pl. cxxxvi. figs. 10-21. The form of the leaflet, contracted near the apex into a short obtuse acumen, is different from any of the European species. A fragment only of a seed referable to this genus has been found, probably at the same locality, being labeled Middle Park, a name often used for leaves from Florissant.

Hab.—Florissant. U. S. Geol. Expl. Dr. *F. V. Hayden.*

Podogonium Americanum, Lesqx.

U. S. Geol. Rep.," vii, p. 298, pl. lix, fig. 5; lxiii, fig 2; lxv, fig. 6.

CASSIA, Linn.

Cassia Fischeri, Heer.

"Fl. Tert. Helv.," iii, p. 119, pl. cxxxvii, figs. 62-65.

Leaflets membranaceous, petioled, ovate-lanceolate, acuminate; secondary nerves at an acute angle of divergence.

These leaves, with the shape, size, and nervation of this species, are acuminate, like fig. 64 of Heer.

Hab.—Florissant. Lacoe's Collection, No. 42.

LEGUMINOSITES.

Leguminosites serrulatus, sp. nov.

Plate XXXIX. Figs 7, 8.

Leaves trifoliate, long-petioled, membranaceous; leaflets narrowly lanceolate, sessile, and serrulate; secondary nerves obsolete.

The leaflets are long and narrow, the lateral a little shorter than the terminal, largest in the middle, tapering upward, acuminate or pointed and gradually narrowed to the base. The relationship of these leaves is unknown to me.

Hab.—Florissant. Princeton Museum, Nos. 784 and 785.

Leguminosites alternans, Lesqx.

Hayden's "Ann. Rep.," 1874, p. 315.

Leguminosites cassioides, Losqx.

"U. S Geol. Rep.," vii, p. 300, pl. lix, figs. 1-4.

Leguminosites species.

Plate XXXIX, Figs. 16, 17.

Pistillate ovaries and stamens of *Leguminosæ.*
Hab.—Florissant.   Seen in divers collections.

ACACIA, Neck.

Acacia septentrionalis, Losqx.

Plate XXXIX, Fig. 15 (15*a* enlarged).

"U. S. Geol. Rep.," vii, p. 29), pl. lix, fig. 9 (9*a* enlarged).

MIMOSITES, Lesqx.

Mimosites linearifolius, Losqx.

Plate XXXVII, Figs. 10-13.

"U. S. Geol. Rep.," vii, p. 300, pl lix, fig. 7.

INCERTÆ SEDIS.

Antholithes obtusilobus.

Plate XXXII. Fig. 20

A monosepalous funnel-shaped perianth, cut to the middle in broad obtuse lobes, attached to the ovary; substance hard, membranaceous.
Hab.—Florissant.   Princeton Museum, No. 856.

Antholithes amœnus, sp. nov.

Plate XXXIV, Figs 13-15.

A six-petaloid perianth, apparently monœcious, with six stamens and one pistil distinctly preserved.
Hab.—Florissant.   U. S. Geol. Expl.   Dr. *F. V. Hayden.*

Antholithes improbus, sp. nov.

Plate XL. Figs. 20, 21.

Whorls of four coriaceous segments, open or reflexed, attached by a narrow base enlarged upward, fan-like and undulate-lobed on the borders.

These fragments might represent reflexed scales of conifers but the axis is too narrow. They are comparable to what Heer has named *Equisetum tunicatum*, "Fl. Tert. Helv.," p. 44, pl. xiv, fig. 10, which represents a broken sheath of *Equisetum*.

Hab.—Randolph Co., Wyoming.   U. S. Geol. Expl. Dr. *F. V. Hayden*.

Carpites gemmaceus, sp. nov.

Plate XL, Fig. 19.

Fruits or buds oval, obtuse, short-pediceled in three at the top of a small branchlet. They are striate in the length, like unopened buds of flowers.

Hab.—Florissant.   Princeton Museum, No. 854.

Carpites Milioides, sp. nov.

Plate XL, Fig. 18.

Seeds on slender pedicels, diffusely panicled, oval, thinly striate lengthwise, 3 millimeters long, 2 broad.

Resembles a panicel of *Milium effusum*, Linn.   The seeds are flattened.

Hab.—Florissant.   Princeton Museum.

GENERAL REMARKS.

The number of species enumerated and described from this group is 228; of these Florissant has the largest number (152), while from the Green River Station 24 species only have been determined from specimens obtained in a cut of the railroad just above the station, and which, of course, represent the Flora of the Green River Group. Of the other localities, I have found 15 species in the specimens from Elko, 14 in those from Randolph County, Wyoming, 7 in those from Alkali Station, 6 in those obtained near the mouth of the White River, and of the other localities marked in the table two or three only in each.

With these materials it is not well possible to determine, from a comparison of the plants of each place, the degree of relation of the local vegetable groups, and, therefore, a table of distribution does not seem of great value for that purpose. It is, however, important to record the data, which may help to trace the march of the vegetation on the American continent during the Tertiary; to see also if the different localities, which I formerly referred to the same stage, show traces of identity in the characters of their plants and at the same time to fix, if possible, the age of the very interesting vegetable group of Florissant by its affinity with some local Flora of Europe. And as this volume is, most probably, the last which I shall have opportunity to prepare on Tertiary plants of Western America, I think proper to leave all the materials which have been examined thus far, exposed as clearly as possible for future comparison.

205

TABLE OF DISTRIBUTION OF THE PLANTS OF THE GREEN RIVER AND WHITE RIVER GROUPS.

NAMES OF SPECIES.	Florissant, Elko, Henry's Fork	Green River Station, White River, Randolph County	Alkali Station, Rock Creek	Sage Creek, Barrel's Springs	AMERICAN. EOCENE.	AMERICAN. MIOCENE. Greenland and Arctic, Alaska, Carbon, Union Group, etc.	EUROPEAN. OLIGOCENE. Gypsum of Aix	EUROPEAN. OLIGOCENE. Häring and Sotzka	EUROPEAN. OLIGOCENE. Bornstädt	EUROPEAN. MIOCENE. Fl. Helv.	EUROPEAN. MIOCENE. Œningen	EUROPEAN. MIOCENE. Bilin	Recent
CRYPTOGAMÆ.													
FUNGI.													
Sphæria Myricæ, Lx		G. R											
CHARACEÆ.													
Chara ? glomerata, Lx	Fl												
MUSCI.													
Fontinalis pristina, Lx	Fl												
Hypnum Haydenii, Lx	Fl												
RHIZOCARPEÆ.													
Salvinia cyclophylla, Lx	Fl											Rel	
Salvinia Alleni, Lx	Fl										Rel		
LYCOPODIACEÆ.													
Lycopodium prominens, Lx	Fl												
EQUISETACEÆ.													
Equisetum Haydenii, Lx			B. S.										
Equisetum Wyomingense, Lx		G. R											
ISOETEÆ.													
Isoetes brevifolia, Lx	Fl							Rel					
FILICES.													
Sphenopteris Guyotti, Lx	Fl				Spitz. Rel								
Adiantites gracillimus, Lx	Fl							Sotz.					
Lastræa (Goniopteris) intermedia, Lx	H. F.										Id?	Id?	
Pteris pseudo-pennæformis, Lx	H. F										Id?	Id?	
Diplazium Muelleri, Hr	H. F							Id					
Lygodium neuropteroides, Lx			B. S.										
Lygodium Dentoni, Lx		W. R								Rel			
CONIFERÆ.													
Pinus Florissanti, Lx	Fl												
Pinus palæostrobus ?, Ett	Fl							Häi.					
Sequoia affinis, Lx	Fl									Rel		Mioc. Rel	
Sequoia angustifolia, Lx	El					Al., Id.							
Sequoia Heerii, Lx				S. C									
Sequoia Langsdorfii, Brgt	Fl					A.,Car.,Id,					Id		
Taxodium distichum miocen., Hr	Fl					Car., Id,					Id.		
Widdringtonia linguæfolia, Lx	Fl							Rel					
Thuya Garmani, Lx	El												
Glyptostrobus Ungeri ?, Hr	Fl					Id		Rel			Id..	Id	
Podocarpus eocenica ?, Hr	Fl							Id					

www.ingramcontent.com/pod-product-compliance
Lightning Source LLC
Chambersburg PA
CBHW030326270326
41926CB00010B/1524